D. MARSHALL.

RADIANT BASICS

A Basic Course For Radiant Panel Heating Systems

RADIANT BASICS

A Basic Course For Radiant Panel Heating Systems

Authored by John Siegenthaler, P.E.

Contributions by Hoyt Corbett, Lawrence Drake, John Fantauzzi

for the

RADIANT PANEL ASSOCIATION
P.O. BOX 717
LOVELAND, CO 80539

An official training manual of the Radiant Panel Association

published by
Teal International Corporation
Loveland, CO 80539

The Radiant Panel Association is made up of members from all levels of the industry who have a stake in the future of radiant panel heating and cooling. We want to insure that your efforts are not only profitable, but provide the public with the best products and services possible. This means a concerted effort towards education, high standards, and open communication. We invite you to join us!

Membership

Companies join the RPA in one of five categories, with each company employee having membership rights:

- Radiant Equipment Supplier
- Associated Equipment Supplier
- Distributor/Wholesaler
- Dealer/Contractor
- Trade Associate/Sales Rep/Professional

Organization

The RPA Board of Directors are elected from the membership by the membership. Various committees are established as needed and any RPA member is eligible to be a committee member. Committee meetings are open to everyone. The membership meets twice a year with the annual business meeting held at the RPA Conference and Trade Show each Spring. Independant affiliated RPA chapters are organized on a local and regional level.

Radiant Basics-

Prepared by: Radiant Panel Association, PO Box 717, Loveland, Colorado 80539

Library of Congress Cataloging-in-Publication Data

Siegenthaler, John.
Radiant basics : a basic course for radiant panel heating systems / authored by John Siegenthaler ; contributions by Hoyt Corbett, Lawrence Drake, John Fantauzzi for the Radiant Panel Association.
p. cm
"An official training manual of the Radiant Panel Association."
ISBN 1-932137-00-9 (pbk.)
1. Radiant heating. I. Title.
TH7421 .S54 2002
697'.72—dc21

2002012380

contents

contents

contents

illustrations

Section 1 **Introduction to Radiant Heating**

Section 2 **Example Systems**

illustrations

Section 2 (cont'd) **Example Systems**

Section 3 **Hydronic Radiant Panel Systems**

Section 4 **Components and Installation Methods**

illustrations

Section 4 (cont'd) Components and Installation Methods

illustrations

Section 4 (cont'd) **Components and Installation Methods**

Section 5 **Electric Radiant Panel Systems**

Section 6 **Cooling and Air Quality**

Section 7 **Panel Covering**

Section 8 **Design Concepts**

illustrations

The Radiant Panel Association gratefully acknowledges the following contributors: Electro Plastics Inc.: product application drawings on pages 145 and 202; National Radiant Design Center: product application drawings on pages 189, 191-196; Mestek: high-velocity system on page 149; Nu-Tech Industries: ventilator drawings on pages 148 and 150; Siebe Redec: product application drawing on page 199; Watts Radiant: product application picture on page 201.

foreword

Radiant Basics

A Basic Course For Radiant Panel Heating Systems

Preface

You are about to participate in a technical training program dedicated solely to radiant panel heating. This program was developed, and is presented, by the Radiant Panel Association. It will provide you with generic, state-of-the-art information about all types of radiant panel heating technologies. As you'll soon discover, the opportunities are almost unlimited.

As more an more people are discovering, radiant panel heating is unsurpassed in providing true thermal comfort in all types of buildings. For those having experienced radiant heating, the difference in comfort is usually nothing short of remarkable in comparison to other forms of heating.

Market

Radiant panel heating can eliminate the cold feet / warm head syndrome that plagues many American homes in Winter. In fact, it's been said that radiant heating can almost make you forget it's Winter outside as you walk in the door.

Superior comfort as well as a host of other benefits has made radiant panel heating one of the fastest growing segments of the HVAC market over the last decade. At present, the radiant panel heating industry is growing at a rate of 25-30% per year. As evidence, witness the shipments of tubing used in hydronic (water-based) radiant heating systems over the last few years. Forecasts suggest this growth will continue for the foreseeable future.

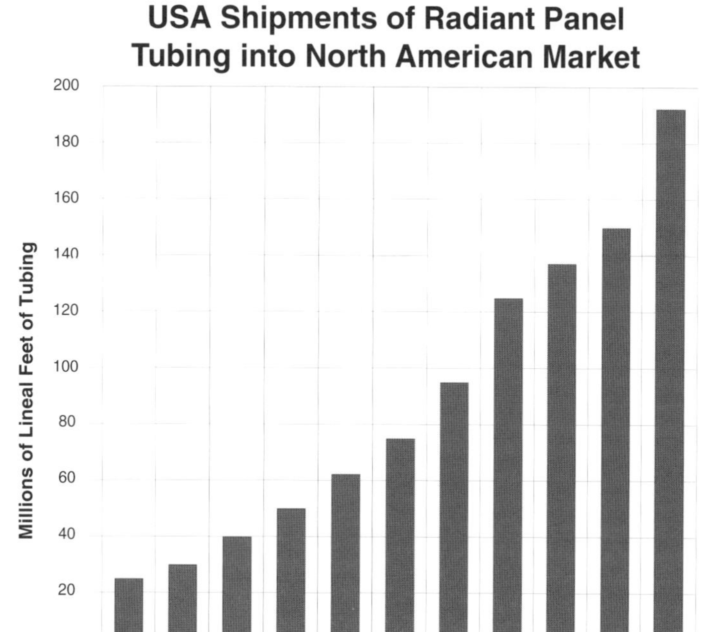

foreword

Opportunities

Your opportunities as a professional installer of radiant panel heating systems has never been greater. To avail yourself of such opportunities, you'll need to understand not only how radiant panel heating works, but also how to design and install quality systems. Many radiant panel systems are literally a part of the building they heat, rather than just hardware fastened to it. Not only must they provide true thermal comfort, they also have to last for may years - preferably as long as the building itself.

Designing and installing a system with such capabilities requires a knowledge of materials and building construction beyond that required for other types of heating. In addition to showing you what materials and system design methods are available, this program also show what pitfalls to avoid.

You'll be exposed to a large amount of information during this training session. Some of it may be familiar to your. A great deal of it, however, has never been published before. This information was gathered from years of experience by many individuals and companies. Collectively, it prepares you to install the ultimate building comfort system.

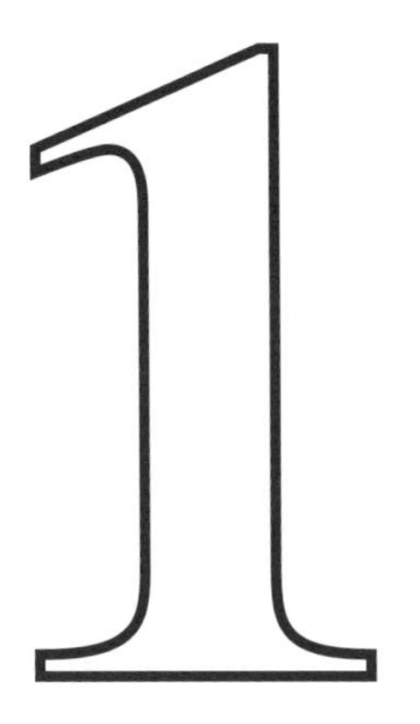

Introduction to Radiant Heating

1•1 What is Radiant Heating?

Radiant heating is a process by which energy leaves the surface of an object and travels to the surface of another (cooler) object in the form of electromagnetic waves.

In some cases this radiation can be seen. Sunlight, for example, delivers a portion of its energy to the earth as visible light.

In other cases the radiation is invisible. Radiant energy, having wavelengths longer than our eyes can see, is called infrared radiation or thermal radiation. This is the type of radiant energy most radiant panels emit.

> Radiant heating is a process by which energy leaves the surface of an object and travels to the surface of another (cooler) object in the form of electromagnetic waves.

All electromagnetic radiation, whether visible or invisible, shares several characteristics:

1. All radiation travels from warmer surfaces that emit it, to cooler surfaces that absorb it.

Heat *always* moves from warmer to cooler materials regardless of what mode of heat transfer carries it.

2. Radiation only becomes *heat* when absorbed by a material.

Sunlight travels through 93 Million miles of empty and frigid space, becoming heat only when absorbed by materials such as gas molecules in the atmosphere or the surface of a driveway. The instant it's absorbed it ceases to exist as electromagnetic radiation. The energy it contained passes to the absorbing material in the form of heat.

3. Any surface can emit radiation to any other cooler surface within sight of it.

The *rate* of radiant energy transfer depends on several factors, including:

a. The difference in temperature between the surfaces
b. The distance between the surfaces
c. The angle between the two surfaces
d. The optical properties of the surfaces (emissivity and absorptivity)

1

4. Radiant heat travels equally well in *any* direction.

Most people think that heat rises. They're actually describing the fact that warm air rises because of its lower density. The fact that radiation travels downward just as effectively as upward is what enables a heated ceiling to effectively deliver heat to objects in the room below.

5. Air absorbs very little thermal radiation.

This is an extremely important (and beneficial) characteristic of radiant heat. It's what allows radiant panel heating systems to directly warm objects without first heating the air around them. It explains why a person can feel the warmth of a campfire even though the air between them and the fire is cold. The radiant energy emitted by the flames only becomes "sensed" as heat when it arrives at and is absorbed by skin and clothing.

6. Thermal radiation can be partially reflected by certain surfaces.

Some surfaces, such as polished metal, can reflect part of the infrared radiation that strikes them just like a mirror reflects visible light. Reflective insulation with one or more bright metallic foil surfaces is designed for just such a purpose. Most common interior building surfaces, however, have low reflective characteristics and therefore absorb most of the infrared radiation that strikes them.

7. Objects with surface temperatures below approximately 970 °F give off infrared radiation that's invisible to humans.

All the radiant panel heating systems described in this training program fall into the infrared category. However, just because radiation emitted from low-temperature radiant panels is invisible doesn't mean it's any less capable of keeping us comfortable. Quite the contrary—in many cases the steady comfort delivered by a low temperature radiant panel is actually more comfortable than bursts of high intensity heat from a high temperature radiant panel. That being said, it's certainly possible to create a radiant panel that gives off a visible glow as it operates. A metal plate heated well above 1000 °F in a blacksmith's furnace is an example. The radiant energy given off by even a small object at such high temperatures can be felt several feet away.

8. All electromagnetic radiation travels at the speed of light.

Regardless of whether it's thermal radiation or visible light, electromagnetic radiation travels away from the surface emitting it at approximately 186,000 miles per second. In the context of sending radiation across even large rooms, this is essentially instantaneous.

9. Thermal radiation is entirely different from nuclear radiation.

Radiant heating systems do not emit subatomic particles or in any way create the health hazards often associated with emissions of nuclear radiation. While most people know this, remember that the word radiation doesn't often bring pleasant thoughts to mind for many consumers. It's important to draw the distinction.

1

10. Thermal radiation is not responsible for all the heat released by a "radiant" panel.

Convection (and in some cases conduction) also plays a role. A typical "radiant" floor system gives off 50% to 70% of its heat as thermal radiation. The remainder is mostly released by gentle convective air currents. The heat output from a typical radiant ceiling however can be as much as 95% radiant. Because warm air against a heated ceiling doesn't "want" to circulate down into a room, convective transfer is very limited.

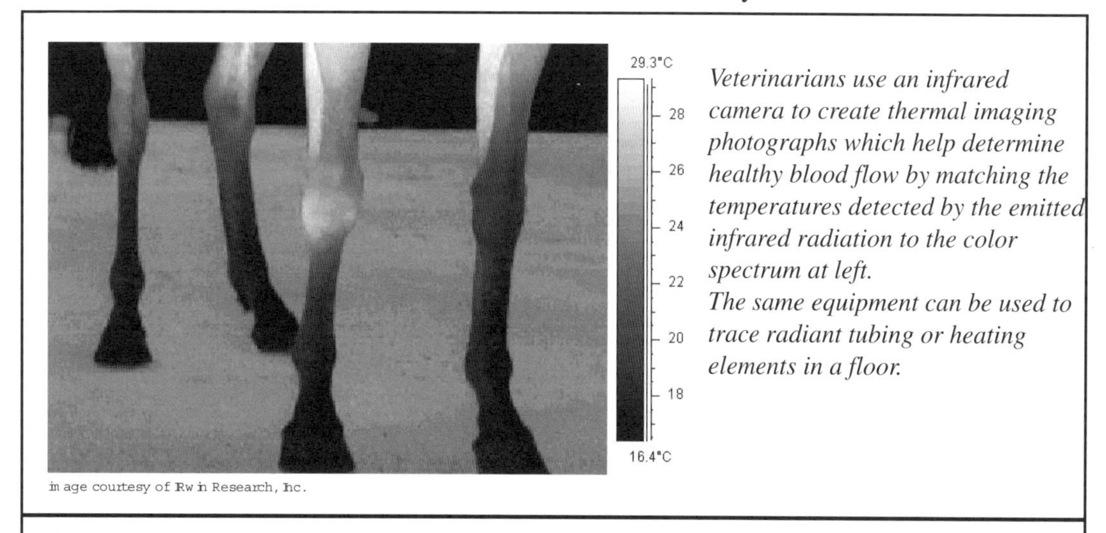

Veterinarians use an infrared camera to create thermal imaging photographs which help determine healthy blood flow by matching the temperatures detected by the emitted infrared radiation to the color spectrum at left.
The same equipment can be used to trace radiant tubing or heating elements in a floor.

image courtesy of Rwin Research, Inc.

Figure 1-1 Surface Temperatures Based on Emitted Thermal Radiation

1•2 Thermal Comfort

To fully understand the advantages of radiant panel heating it's necessary to understand thermal comfort. Although most people know when they're comfortable, few of them know why. "Comfort professionals" need to understand what factors affect thermal comfort and how to create an interior environment that ensures it exists.

Many people think that just because the room temperature is in the range of 70 °F, they should be comfortable. Providing true thermal comfort, however, involves more than simply maintaining inside temperature within a certain range. Comfort is established and maintained only as long as our bodies release the heat generated by metabolism at the same rate it's produced. *When truly comfortable, we should be totally unaware of how or where our bodies are losing heat.* Radiant heat transfer plays a big part in providing—or denying—thermal comfort.

To appreciate the intricacies of maintaining thermal comfort, it's important to understand how our bodies release heat. Under light activity conditions, such as working at a desk, an adult generates about 350 to 400 Btus per hour of heat by metabolism. That's roughly equivalent to the heat given off by a 100 watt light bulb.

Four natural processes are involved in releasing heat from the body:

1

- Evaporation of moisture from the skin
- Heat transferred to the air around the body by convection
- Heat conducted to objects the body is in contact with
- Radiant heat transfer between the body and surrounding objects

The combined effect of all four processes determines if we release heat at the same rate it's generated, and hence whether or not we're comfortable.

Evaporation:

Evaporation of moisture from the skin is responsible for about 25% of the heat output of our body during light activity within typical interior spaces. As the humidity of the air around us increases, evaporative heat loss becomes less effective. Moisture accumulates on our skin faster than it can evaporate. You know the feeling—and the resulting loss of thermal comfort. By contrast, evaporative cooling is very effective (and therefore very noticeable) before a person towels off after a swim, even when the temperature of the air surrounding them would otherwise be considered comfortable.

Convection:

Heat flows from our skin and clothing to the surrounding air whenever the air is at a lower temperature. This is an example of convection heat transfer. As the air in contact with our skin and clothing absorbs heat it gently rises. Cooler air flows in to replace it and continues the process. Under normal interior conditions, while wearing light clothing, convection removes about 30% of the heat our bodies generate. Its effectiveness depends on both the temperature difference between our external surfaces and the air around us, as well as the speed the air flows past us. Increasing either increases convective heat loss. For example, we experience "wind chill" whenever air at a temperature lower than our exterior surfaces flows past us. As the speed of the air increases, we feel a definite increase in its cooling effect even though its temperature hasn't dropped. Although wind chill is usually thought of as an outdoor phenomena, it can also occur inside. The word "draft" is used to describe this (usually undesirable) effect during the heating season. To maintain comfort, heating systems must avoid creating noticeable drafts.

Conduction:

Heat loss by conduction occurs whenever part of our body touches a colder object. If you stand barefoot on a cool basement floor, your feet quickly feel the result of conduction heat loss. The warmed seat of a chair after someone has been sitting in it for a while is also the result of conduction heat loss.

Radiation:

Radiant heat transfer from our skin and clothing to objects around us plays a big role in determining thermal comfort. Whenever we're in the proximity of objects cooler than our exterior surfaces, we radiate heat to them. Under normal interior conditions with light clothing, almost **half of our body's total heat output** is released by radiation. This is what creates the chill we feel while standing near a cool window surface, even though the air around us may be 70 °F or more. Another way to feel radiant heat loss is to walk out from under an overhang on a clear night. There is an instant sense of the increased cooling effect as your body comes into view of the cold upper atmosphere that was previously shielded by the overhang. The surrounding air temperature didn't change, but the increased radiational cooling makes it feel that way.

Our bodies automatically adjust all these processes to regulate heat loss. When one method of heat loss is limited, the body attempts to increase its other heat loss mechanisms. For example, when high humidity inhibits evaporative heat loss, our skin temperature rises in order to boost both convection and radiation heat loss. In such situations our instincts often tell us to increase convection by fanning our face, standing where there's a breeze, and so forth. But there are limits to how far our bodies can adjust one or more of these mechanisms to compensate for others. That's where a building's "comfort system" comes in. Ideally, it works in combination with our body's own heat loss mechanisms, keeping them within a fairly narrow range so they can "fine-tune" body heat output and thus maintain comfort.

We've probably all experienced most of the uncomfortable situations described above, but few of us have paused to give them much thought. Now that you have a basic understanding of how these natural processes work, you'll become more discerning of your thermal surroundings and the degree of comfort they offer. Remember, thermal comfort is achieved when we are totally unaware of how or where our bodies are losing heat.

1•3 Advantages of Radiant Panel Heating

A properly designed and installed radiant panel heating system offers many advantages. Some are simply the elimination of a "side effect" created by another method of heating. In other cases, they're unique conditions that only radiant panel heating can provide. Here's a listing of the major benefits along with a brief explanation of each.

1. Radiant panel heating improves comfort by increasing a room's average surface temperature.

Remember that a large portion of our body's heat loss is by radiation to cooler surfaces surrounding us. The cooler these surfaces are, the faster they "pull" heat from us and the more uncomfortable we feel. By warming the interior surfaces of floors, walls, ceilings, windows and doors, radiant panel heating reduces radiant heat loss from our bodies. Because we're especially sensitive to our radiant surroundings, this significantly improves comfort.

2. Radiant panel heating allows comfort at lower air temperatures.

Because radiant heat loss from the body is reduced, a greater portion of the body's heat output shifts to convection. This in turn necessitates a greater temperature difference between our exterior surfaces and the air around us. Most people engaged in light activity will be very com-

1

fortable in radiantly heated rooms with air temperatures in the mid 60 °F range. If the activity level is greater, light shop work for example, comfort is often maintained with air temperatures in the low 60 °F range.

3. Radiant (floor) heating provides an almost ideal match to human thermal comfort requirements.

The "ideal" temperature variation from foot to head level is shown in Figure 1-2 along with the "typical" temperature profile created by other heat delivery systems. Note that the temperature profile created by the heated floor is an almost perfect match to the ideal scenario.

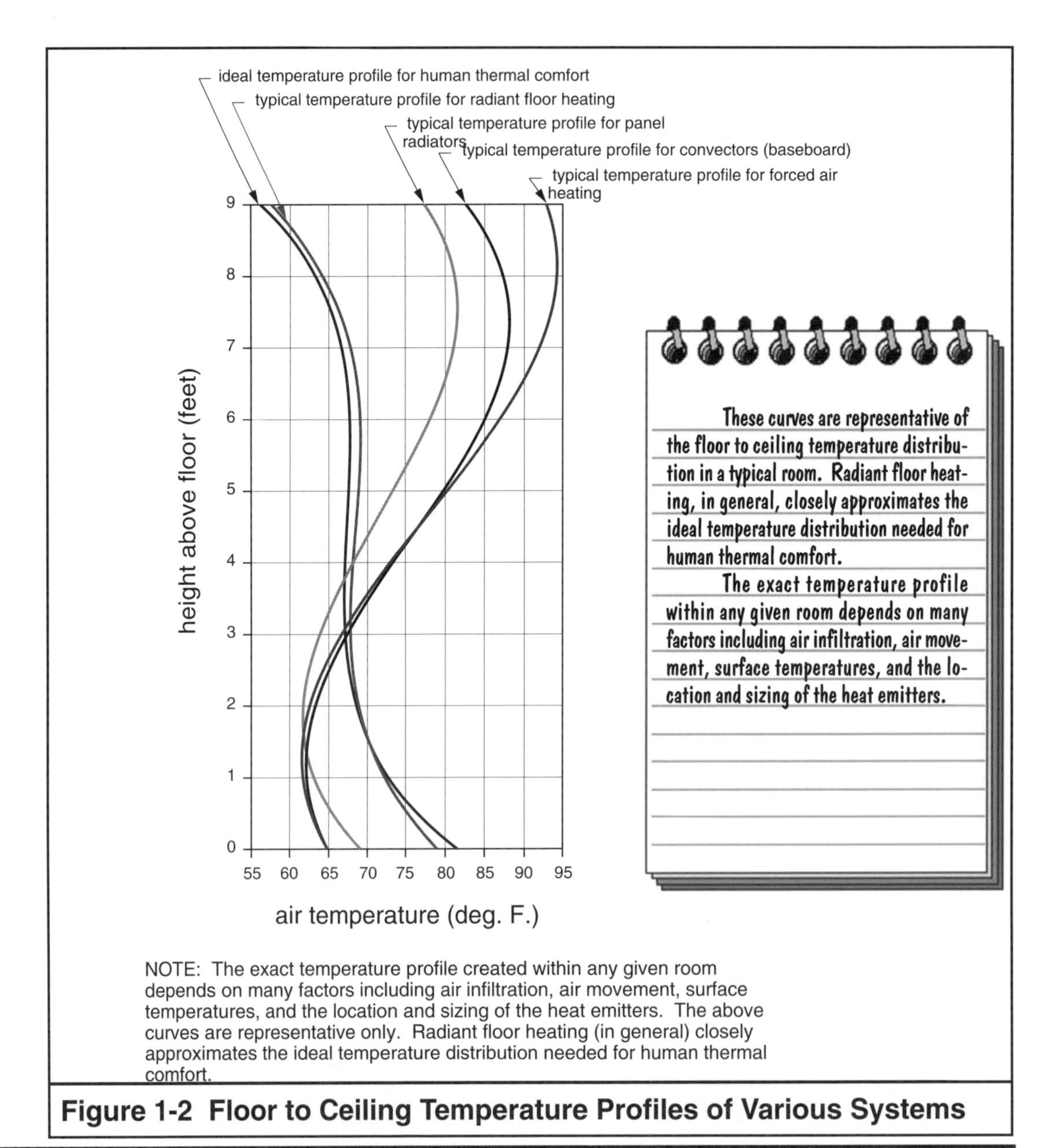

Figure 1-2 Floor to Ceiling Temperature Profiles of Various Systems

4. Radiant (floor) heating reduces or in some cases eliminates room temperature stratification.

The familiar (but uncomfortable) situation of warm air blanketing the ceiling while cool air pools at floor level is eliminated. Not only does this improve comfort, it reduces heat transfer through the ceiling and upper portions of the wall. It also reduces "stack-effect" air leakage from the building. The greater the height of the interior space, the more beneficial this characteristic is. The need for "paddle fans" to push warm air back down to the occupied zone of the room is eliminated in most cases.

5. Many radiant panel heating systems are totally out of sight.

Whether constructed with hydronic tubing or electric cable, heated floors, walls, and ceilings conceal their function as heat emitters. There are usually less constraints imposed on furniture placement. The architectural aesthetics of rooms can be preserved without compromising thermal comfort.

6. Radiant heating systems are easily zoned.

Many radiant panel heating systems, both hydronic and electric, are designed for room-by-room zoning. Not only does this offer the potential for energy savings by reducing temperature in unoccupied rooms, it also allows different occupants to adjust rooms to their own desired comfort level. In addition, room-by-room zoning helps prevent localized overheating due to solar or other types of internal heat gain.

7. Radiant heating systems create gentle *room* air circulation.

In most cases the air motion is so slow that it's not detectable by occupants. Central forced-air systems, by contrast, often create very noticeable whole-building air circulation. The latter tends to distribute dust, odors, and airborne viruses throughout the entire building. Some central forced-air systems also pressurize rooms as they operate, increasing air leakage and wasting energy.

8. Hydronic and electric distribution systems supplying radiant panels are easily and unobtrusively routed through buildings.

The small tubing or electrical cables supplying radiant heating panels are easily concealed within (or routed through) the structural cavities of most buildings. By contrast, air ducts of equivalent heat carrying ability often require careful planning and structural alterations if they're to be totally concealed. (Figure 1-3)

9. Radiant heat systems with high thermal mass (such as heated concrete slabs) can respond quickly to increased loads when necessary.

Many heat delivery systems such as forced air and hydronic baseboard have low thermal mass. This limits the rate of heat output to the heating capacity of their heat source. By contrast, a heated floor slab with its large reservoir of heat can release a "burst" of heat when necessary. An example is when a large overhead door in a garage is opened and cold air floods in across the floor. Comfort conditions are

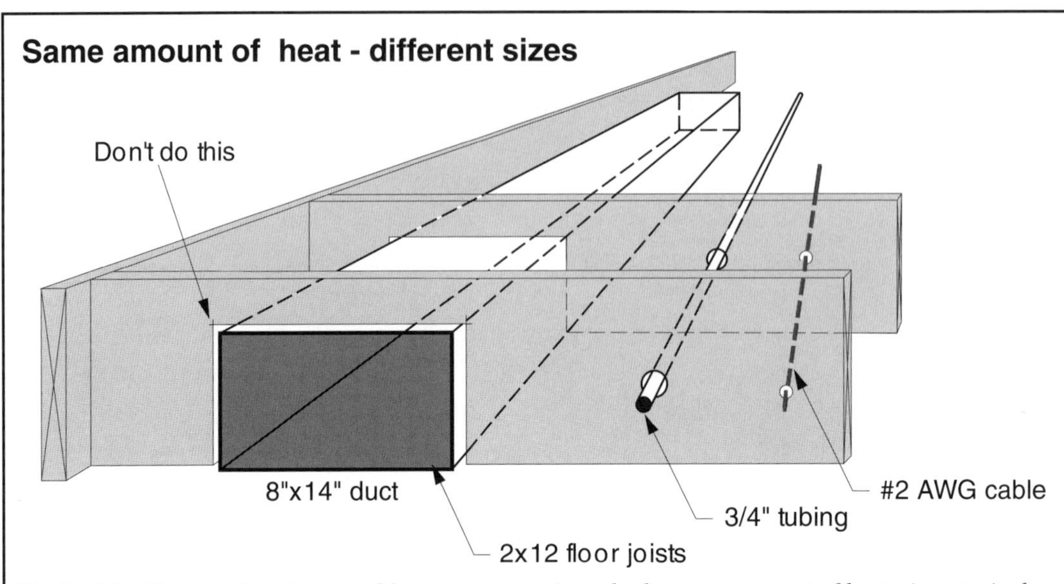

Each of the illustrated systems could convey approximately the same amount of heat given typical operating conditions. The tubing and cable are easily routed through the framing. The hypothetical cut shown in the 2x12 floor joist to accomodate the duct would actually destroy its structural integrity.

Figure 1-3 Heat Transport Capacity of Duct, Tube, and Electric Cable

reestablished very quickly after the door is closed.

10. Radiant heating systems with low thermal mass (such as electric ceiling panels) release heat almost instantly when turned on.

This characteristic allows an almost instantaneous improvement in the thermal comfort of a cool room. Although the objects and air in the room will require some time to warm, radiant heat loss from the body is reduced within a couple of minutes of turning on the low mass panel.

11. Heated floors dry quickly.

Not only does this improve safety, it also provides much better working conditions for those who routinely work on or near wet floors. It's a tremendous benefit in facilities such as auto service garages, fire stations, or other vehicle service facilities.

12. Many types of radiant panels are very resistant to physical damage.

Tubing or wiring encased in concrete floors is very resistant to damage from heavy interior traffic. Likewise, heated ceilings are typically "above the fray" and thus less likely to be damaged by normal building usage.

13. Radiant heating panels can operate with virtually no noise.

Properly designed and installed systems eliminate expansion noise, sheet metal vibration, or velocity noises caused by fast moving air. However, attention to detail during installation is crucial in achieving a quiet distribution system.

14. Low temperature hydronic radiant panels are adaptable to almost any type of heat source and fuel.

Low-temperature heat from solar collectors, geothermal heat pumps, or waste heat recovery systems can often be utilized. Such heat is only marginally suitable for other types of delivery systems.

15. Radiant panel heating reduces energy consumption.

There are several reasons that support this statement:

- Since comfort is maintained at lower air temperatures the heating load associated with air exchange is reduced. The greater the heating load due to air exchange, whether by natural leakage or forced ventilation, the greater the energy savings.

- Easy room-by-room zoning control allows unoccupied spaces to be reduced in temperature, thus reducing total heating load.

- Reduced rates of air leakage by eliminating the "stack effect" inside buildings.

- Reduced conductive heat loss through ceilings by eliminating the layer of warm air normally present against its lower surface.

- Low water temperatures in some types of hydronic radiant panel systems improve boiler efficiency and reduce heat loss from distribution piping.

- High thermal mass systems have the potential to make good use of low cost "off-peak" electrical utility rates where available.

- The electrical energy consumption of small circulators used in hydronic radiant panel systems is usually lower than blowers in forced-air systems of similar heating capacity.

Estimates of how much energy a radiant panel heating system saves relative to other methods of heating can vary considerably. It's not uncommon to hear savings claims of 20 to 30% or more. Many factors influence the extent of energy savings attained by a given system. For example, how well will a given project "exploit" the energy saving characteristics noted above. Buildings with high ceilings, relatively poor insulation, high air exchange rates, or a combination of these characteristics definitely have the potential for greater savings using a radiant panel heating system. Certainly the quality of design and installation of the radiant panel system as well as the system it's being compared to will affect the results.

Heating professionals are urged to use caution in stating energy savings to prospective clients. In time, each company installing radiant panel systems has the opportunity to build a portfolio of projects, some of which have a "before and after" record of fuel usage. Such documented projects provide a credible way of demonstrating the energy savings of radiant panel heating. It's also important to keep in mind

1

1•4 A Brief History of Radiant Panel Heating

that although reduced energy usage is likely, and certainly desirable, it's seldom the top priority of today's consumer.

The first recorded use of radiant heating in buildings was by the ancient Romans who used it in their well-known bath houses. These buildings were designed with chambers that allowed fires to be placed under the floors and within walls. The hot exhaust gases were routed through flue-like chambers, giving up their heat to the masonry surfaces along the way. Patrons were not only bathed in warm water, but the radiant heat from the warm walls and floors as well. As crude as these systems were by today's standards, they obviously offered comfort unattainable by simpler means. Their intricate construction testifies that comfort was highly valued, and justified the elaborate construction necessary to achieve it.

The first use of hydronic radiant floor heating using iron pipe was in England around 1907. In the years that followed, the English applied it in all types of buildings, including stores, schools, and hospitals.

Hydronic floor heating gained exposure in the US through the well-known designs of architect Frank Lloyd Wright, among others. Wright understood the need to provide superior comfort as part of the experience of living in a one-of-a-kind home. He used radiant heating to transform the massive stone and masonry surfaces typical of his designs into inviting elements, rather than the harsh cold surfaces they are often perceived as. Concerning radiant heating he wrote: "No heating was visible nor was it felt directly as such. It was really a matter not of heating at all, but an affair of climate."

During the post-war housing boom of the late 1940's, thousand of hydronic floor heating systems were installed in the United States using copper tubing and steel pipe. Although some of these systems are still operating today, many failed due to metal

Figure 1-4 Hydronic Radiant Pipe in the Early 1900's

fatigue or chemical incompatibilities with concrete. When a leak occurred, it was often difficult or impossible to find and correct. Eventually many of these early systems were simply abandoned in favor of other methods of heating. Skepticism understandably developed and interest in hydronic floor heating declined. The "death blow" to the first generation of hydronic floor heating was the advent of central air conditioning in the 1960's. Since a forced-air system could transport both heated air in winter and cooled air in summer, and didn't suffer from the potential leakage problems of early floor heating systems, it quickly became the standard of the American housing industry.

Rapid growth in the US home building industry over the last 40 years spawned a very competitive and price-dominated market for heating systems, a market that willingly substituted lower cost, and faster installation, for proven comfort. Many home buyers were apathetic about the type of heat in their new house. Most just assumed their builder knew the best way to heat their house and accepted his choice without question. Unfortunately, many became "casualties" of an apathetic attitude toward comfort that accompanied the fast-tract building ethic. Many would now be more discriminating about the method of heating used in any new home or addition they build.

A recent survey conducted by an independent consumer research firm found that 80% of the 80,000 American households responding were only somewhat or not at all satisfied with their heating system. Only 16% said that price was the most important factor in purchasing a new heating system. The message is clear: Most people want better comfort from their home's heating system, and most would willingly spend more money to achieve it.

Without a doubt, *comfort* has always been the most sought-after benefit of radiant panel heating. Even early hydronic radiant heating systems, though sometimes plagued by premature material failures, delivered superior comfort for their time. A way was needed to retain this comfort, while at the same time making hydronic floor heating easy to install, and capable of a long, trouble-free life.

During the 1960's, as American interest in floor heating was rapidly declining, research was underway in Europe on a new polymer material called cross-linked polyethylene, or "PEX" for short. Although originally developed as a sheathing for underwater cable, it would become the material that eventually revolutionized the hydronic floor heating market worldwide.

Today's hydronic radiant panel heating systems are as different from their predecessors as compact disks are from phonograph records. Several kinds of high-tech piping materials now make it possible to install systems that can last for decades, perhaps even longer than the buildings they are a part of. These materials have been extensively tested in radiant panel heating applications over the last three decades. Each year, hundreds of millions of feet of tubing are installed in new radiant panel heating systems worldwide. Major advances have also been made in heat sources and controls. This state-of-the-art hardware has effectively merged the superior comfort aspects of radiant heating with market demands for easy installation and reliability.

1

Example Systems

2•1 Introduction

This section presents several contemporary applications of radiant panel heating. The projects represent a broad range of radiant panel technology, including both hydronic and electric systems. Both residential and light commercial applications are shown.

Some of these systems are the essence of simplicity; others use more sophisticated technology where it's appropriate. Each case study includes a brief description of the building and its heating system. Each includes photos and illustrations that bring out key elements of its radiant panel heating system.

After completing later portions of this training program, you're encouraged to return to these case studies and look for the details you've learned about. If a picture is worth a thousand words, there's a lot of information available within these case studies.

2•2 Residential Systems

CASE STUDY 1
Small house with hydronic slab-on-grade floor heating supplied by a water heater.

The small guest house shown in this case study was recently constructed in southwestern Wyoming. Although only 800 square foot in size, it provides comfort that the owners wish they had in their main house. Since the guest house was planned as a slab-on-grade structure the addition of hydronic tubing (and underslab insulation) was simple and economical. The small size of the building, in combination with good insulation, resulted in a very low heating load. Since the guest house would have only one bathroom and its domestic water load relatively small, a propane-fired water heater would be able to provide both domestic hot water as well as space heating.

The system uses several circuits of 1/2" PEX tubing cast into the floor slab. The manifold where each of the tubing circuits begins and ends is connected to the water heater by a bronze circulator. Because none of the components in the heating sub-system are ferrous, domestic water from the tank can be circulated directly through the floor circuits and back to the tank. The water temperature maintained by the DHW tank is suitable for direct circulation through the floor circuits without need for a mixing device. The system's circulator is simply turned on whenever the thermostat calls for heat.

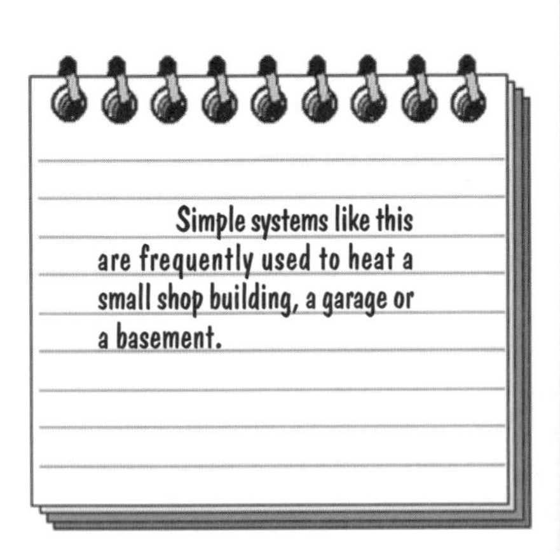

Figure 2-1 A basic 800 sq. ft. slab-on-grade guest house with a simple hydronic radiant floor heating system offers exceptional comfort.

2

Figure 2-2 1/2" PEX tubing with oxygen barrier being embedded in concrete slab floor. Note tubing risers for manifold station near center of slab.

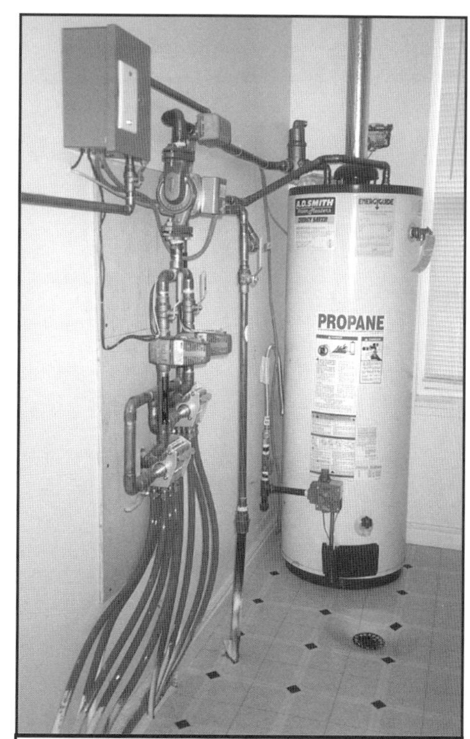

Figure 2-3 A standard water heater supplied heat and hot water to this small building.

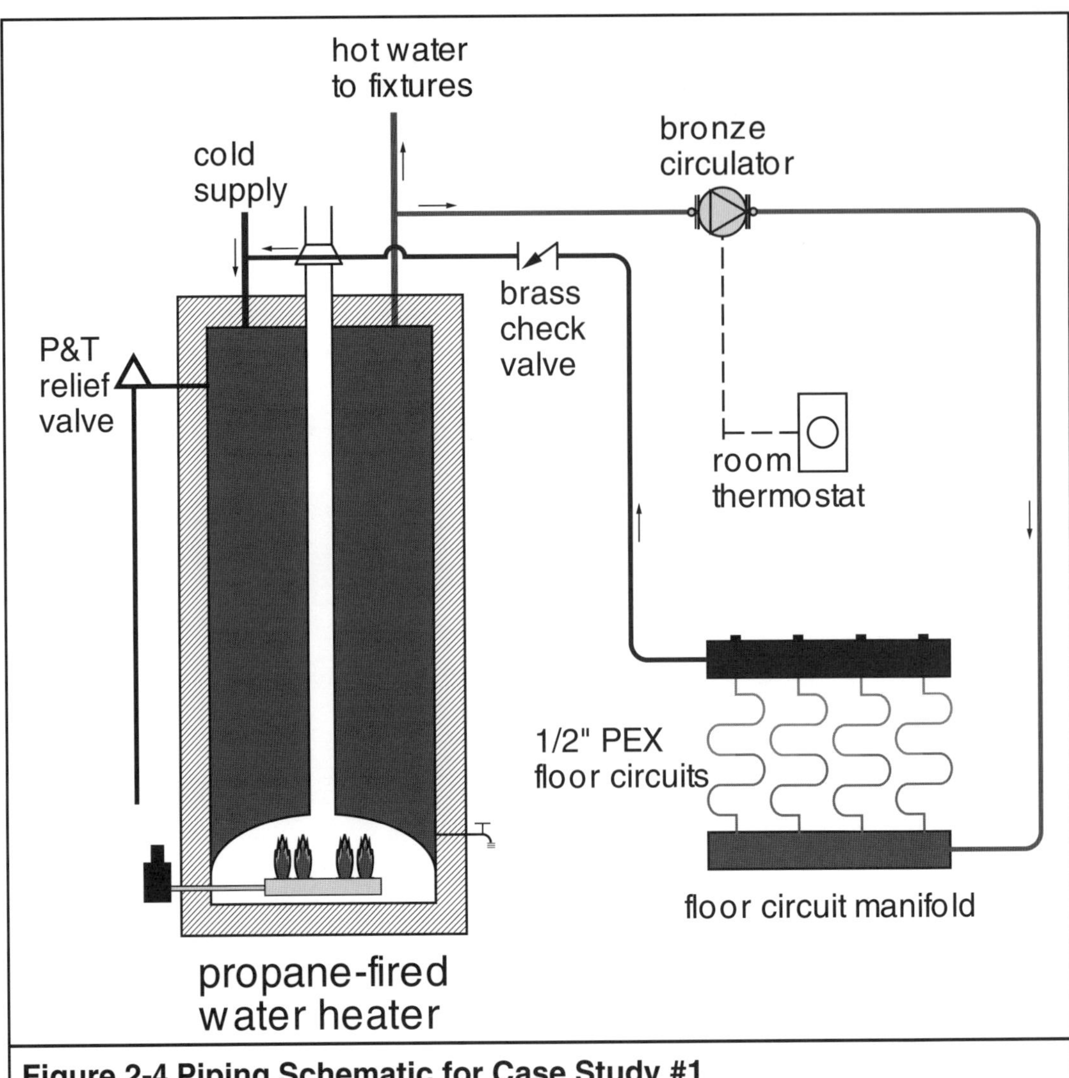

Figure 2-4 Piping Schematic for Case Study #1

In Figure 2-4 the thermostat turns on the bronze circulator whenever the building needs heat. Tank temperature is controlled by its own internal thermostat.

This configuration is considered an "open" system since the domestic hot water is also used for the radiant heating system.

With open systems, all components in the radiant heating system must be non-corrodible and suitable for contact with potable water. The components must be able to withstand the pressure rating of the domestic plumbing system. And, space heating output will be limited during periods of high DHW demand possibly requiring the use of DHW priority controls.

2

CASE STUDY 2
House with hydronic (thin-slab) floor heating and DHW supplied from same boiler.

Located in a severe (-40 °F) winter climate in Maine, this house uses a multiple zone hydronic floor heating system. Each zone is supplied by its own manifold station. One zone serves the basement area, another the garage, and the third serves the entire first floor. Tube spacing was adjusted to allow the same water temperature to supply all the floor circuits. Different tube spacings were required due to the different floor coverings used. This approach reduces the number of mixing devices necessary.

One-half inch PEX tubing was stapled down between 1.5" sleepers on the first floor. The sleepers provided nailing for the hardwood flooring in certain areas of the house. After pressure testing, the space between sleepers was filled with concrete to help spread the heat across the floor area and add thermal mass. The rough opening heights of the windows and doors were adjusted during framing to allow for the thickness of the thin-slab system. Tubing spacing varied from 12" on center in some areas to 6" on center in high heat loss areas.

In the basement and garage the tubing was secured to welded wire reinforcing laid over 2" extruded polystyrene insulation. Tube spacing was 12" o.c. in the finished areas of the basement, and 18" o.c. in the garage and storage areas. A 4" concrete slab was poured over the tubing.

Heat for both space heating and domestic hot water is supplied by a conventional oil-fired boiler connected to a short "primary" piping loop. High temperature water is supplied to an indirect domestic water heater connected as a "secondary" loop. Reduced temperature water for the radiant panels is provided using a variable-speed pump (reverse) injection mixing system. The injection mixing sub-system is also connected as a secondary loop. This piping arrangement allows each load to operate without circulator interference. All three zones use a parallel anticipator low voltage thermostat to control heat input.

Figure 2-5 A -40 °F Location in Maine

Figure 2-6 Tubing and Sleepers Awaiting Concrete Fill

Figure 2-7 Mechanical Room in Basement

2

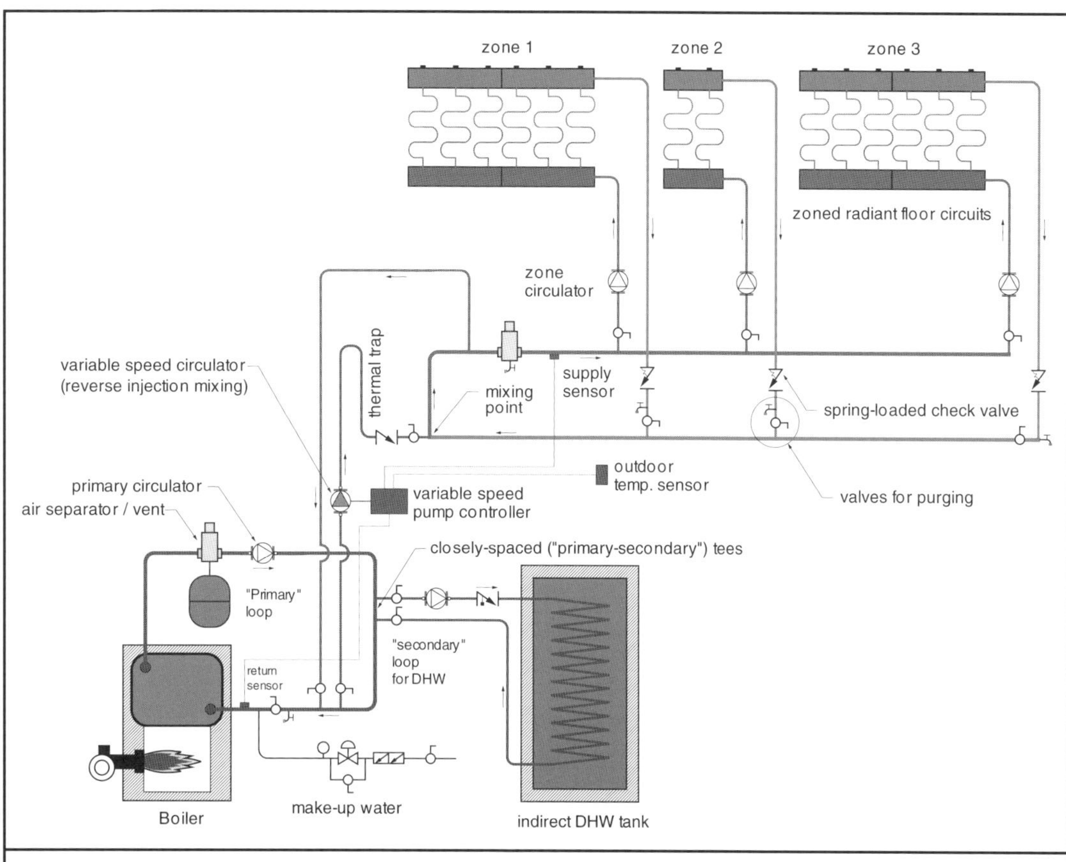

Figure 2-8 Piping Schematic for Case Study #2

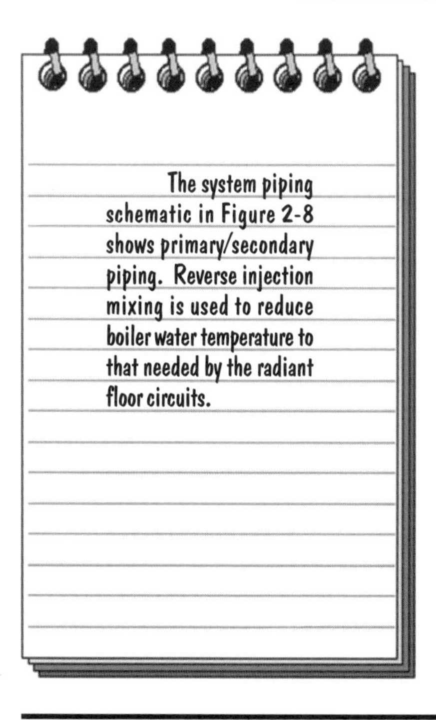

CASE STUDY 3
Private house retrofit with hydronic panel radiator system.

Some hydronic radiant panel systems are well-suited to retrofit applications. The building in this case study is a passive solar house constructed in 1983 in upstate New York. A small wood stove served as the back-up heating system during low solar periods (which are frequent in this location). When the local utility installed gas mains adjacent to the house, the owners, who are retirement age, decided to install a fully automatic heating system.

The system consists of 9 panel radiators installed in the various rooms of the house and piped back to a small mechanical room using 1/2" PEX-AL-PEX tubing. Because the house was framed using open-web floor trusses rather than conventional floor joists, the tubing was easily pulled through the floor deck to several locations and then routed down through partitions (or surface mounted to masonry walls where necessary). The supply and return tubing from each radiator is connected to a manifold system much like that used in a hydronic radiant floor system.

Heat is supplied by a direct-vent gas-fired boiler which eliminated the need to install a new chimney. Hot water from the boiler is reduced in temperature by a variable speed pump (direct) injection mixing system. Water temperature supplied to the panel radiators is controlled by reference to outside temperature (reset control).

Each panel radiator has its own thermostatic control valve allowing each room to be individually zoned. This was a significant benefit given the home's passive solar design. On cold but clear days the panel radiators in rooms with solar gains quickly turn themselves off, while those in north-facing rooms continue to provide heat as required. The low thermal mass of the panel radiators allows fast response to changing load conditions without overheating.

Figure 2-9 Hydronic Panel Radiant Retrofit

Figure 2-10 Panel Radiator with Thermostatic Valve

Figure 2-11 PEX-AL-PEX Routed Through Open-Web Trusses

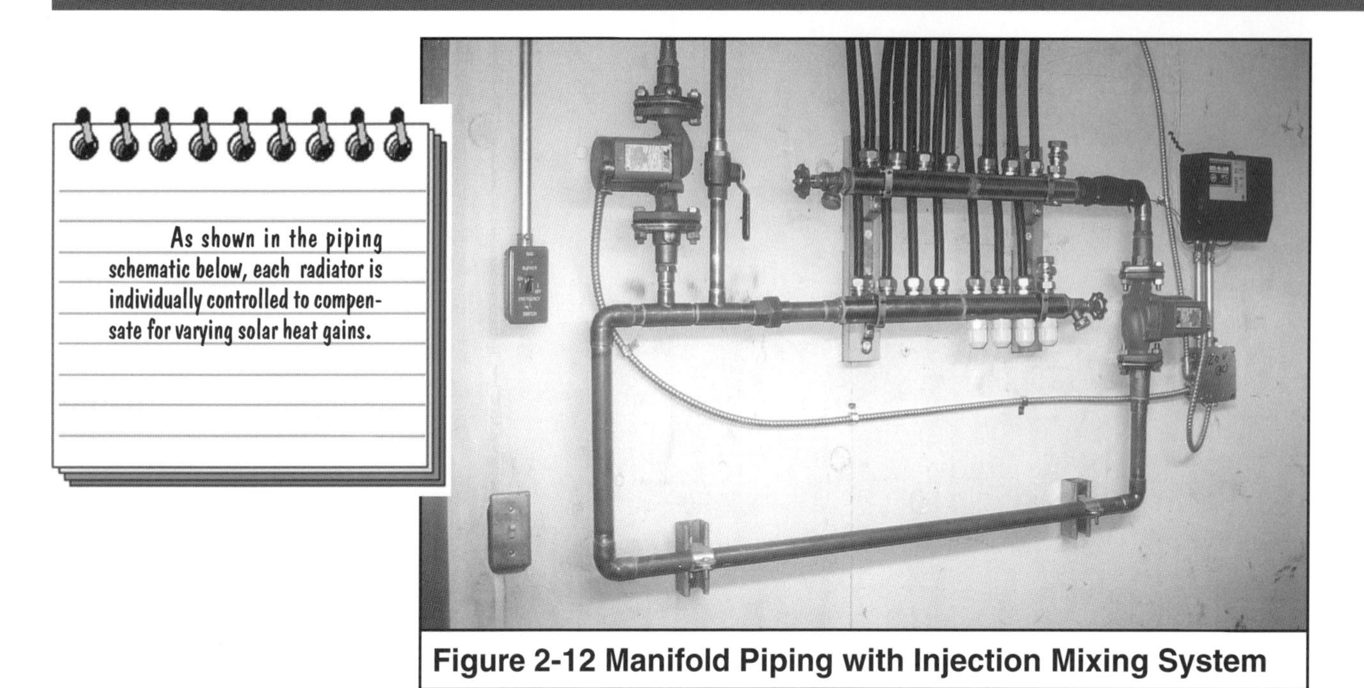

As shown in the piping schematic below, each radiator is individually controlled to compensate for varying solar heat gains.

Figure 2-12 Manifold Piping with Injection Mixing System

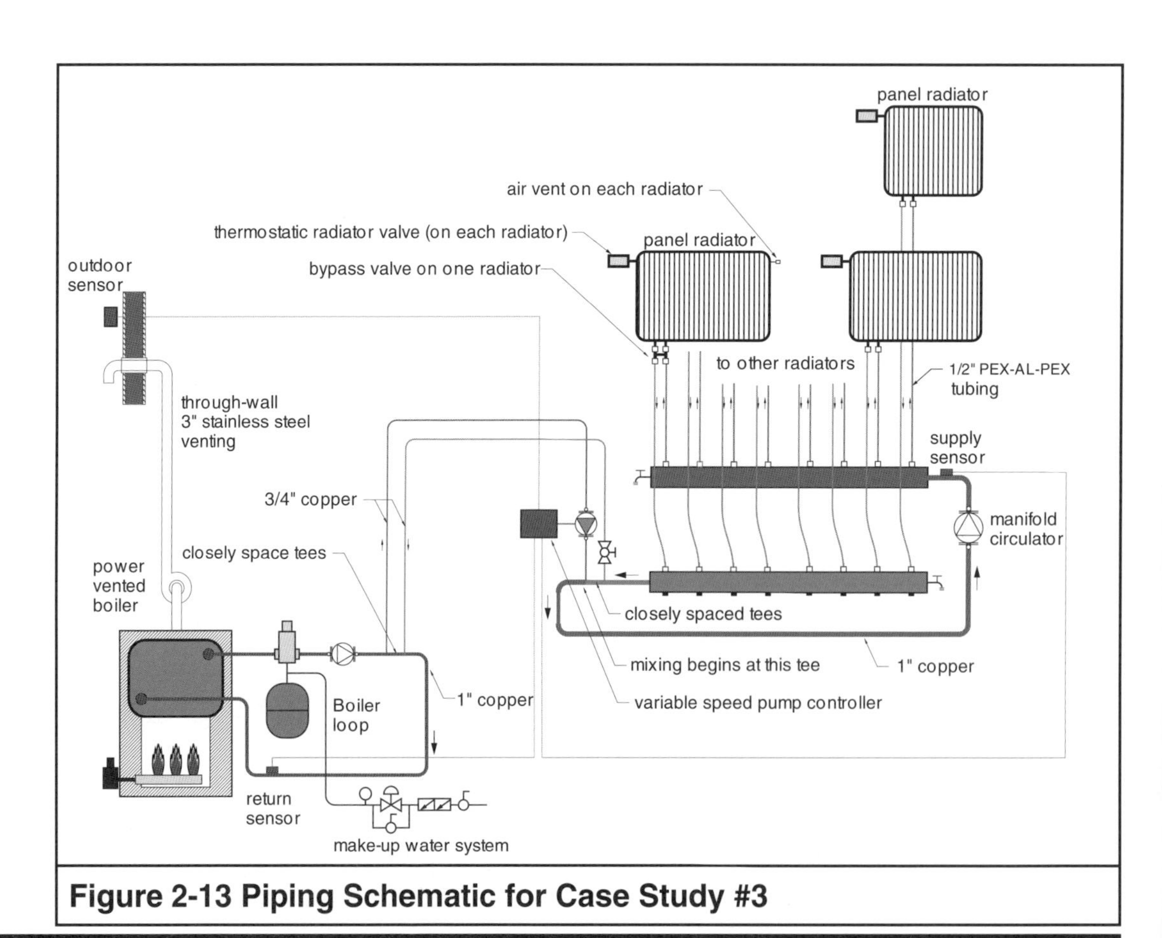

Figure 2-13 Piping Schematic for Case Study #3

2

CASE STUDY 4
Entry foyer heated by an embedded electrical cable system.

This case study demonstrates the use of a small electric radiant floor panel in an otherwise hydronically heated home. Electric heating cable was fastened in place directly over the wood subfloor. It was then embedded in a thin concrete slab. The slab is covered with ceramic tile that offers excellent heat transfer and durability in this high traffic area. The electric system can be completely turned off (if desired) without concern for freezing. A special thermostat which measures both air temperature and slab temperature controls the system. The heated floor rapidly dries moisture tracked in as well as warming boots and clothing stored in the foyer.

Electric systems are often used for whole house heating, area heating and floor warming.

Figure 2-14 Entry to Heated Foyer

Figure 2-15 Foyer Tile Warmed by Electric Cable

2

Figure 2-16 Electric Cable on Wood Subfloor

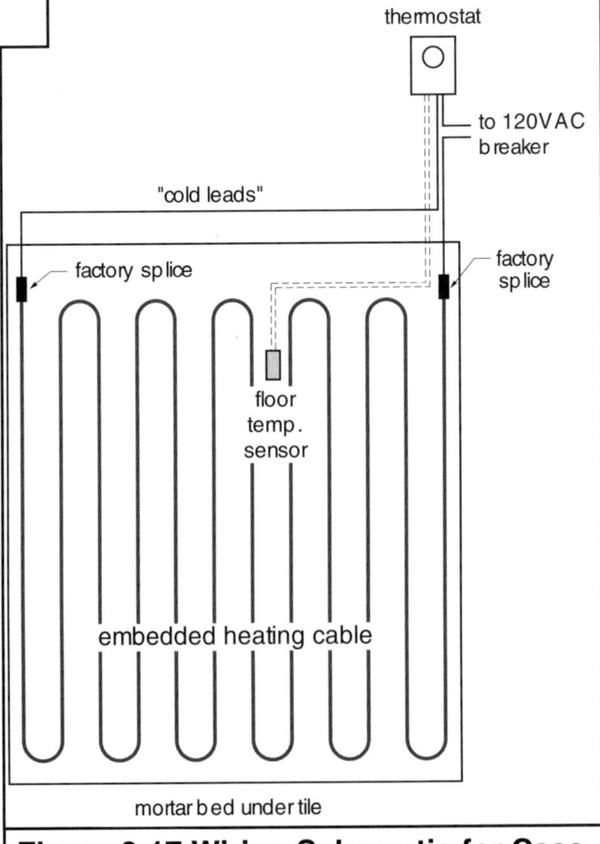

Figure 2-17 Wiring Schematic for Case Study #4

2

CASE STUDY 5
House using open-loop geothermal heat pumps to supply a hydronic radiant floor heating system.

The 3600 square foot house shown in this case study was constructed in central Maine. It uses two water-to-water heat pumps as its heat source. Both heat pumps are supplied with water from a deep (750 ft.) drilled well. The water temperature from the well averages about 50 °F.

The heat pumps maintain the water temperature of an 80 gallon "buffer tank" that separates the heat generation and heat distribution sub-systems. One heat pump turns on when the buffer tank drops from its setpoint of 110 °F down to 105 °F. The second heat pump turns on if the tank drops to 100 °F.

The radiant floor distribution systems draw warm water from the tank as necessary. The buffer tank allows even a single floor circuit to operate independently of the other circuits without short-cycling the heat pump.

The high thermal mass of the floor heating system allows the owner to take advantage of the "off-peak" electrical rates ($0.05 / KWH) offered by the local utility. The heat pumps can be turned off (or run only as necessary) between 7 AM - noon, and 4 - 8 PM.

In addition to the 5 manifold locations serving the underfloor circuits, the systems uses two fan-coil units for supplementary heating of the sunspace and master bedroom.

Figure2-18 Geothermal/Radiant Heats 3600 Sq.Ft. Home

2

Figure 2-19 PEX Tubing Awaiting Concrete Pour

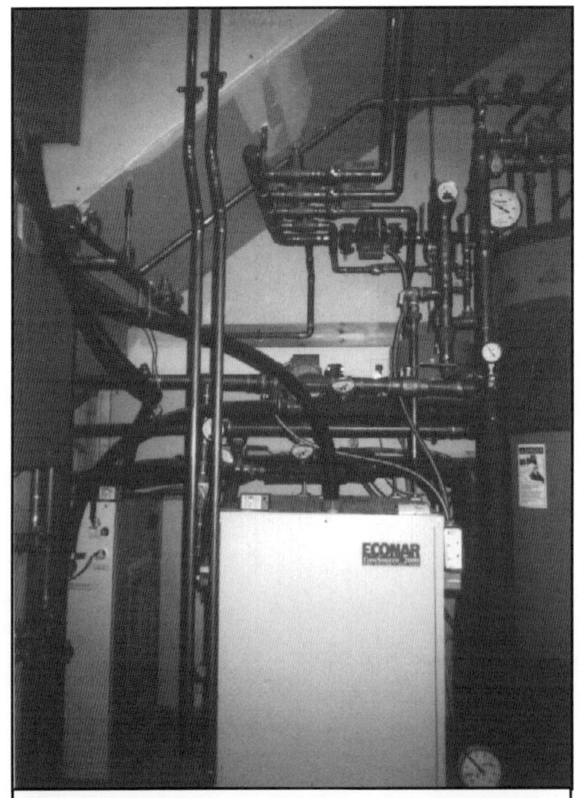

Figure 2-20 Heat Pump with Buffer Tank

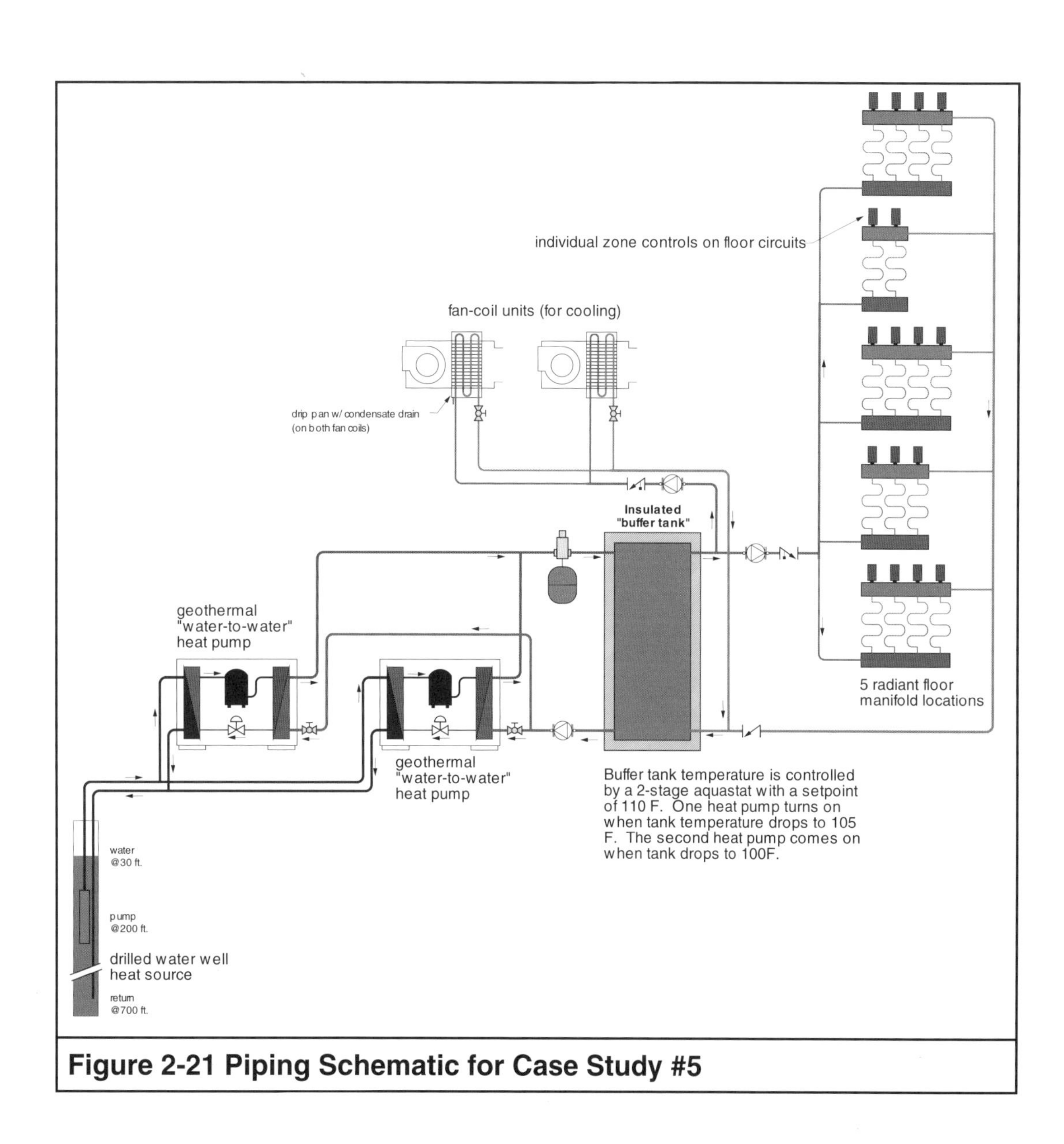

Figure 2-21 Piping Schematic for Case Study #5

CASE STUDY 6

Town municipal building using hydronic floor heating and 4-way mixing valve.

The building shown in this case study is a 7000 sq. ft. single story slab-on-grade structure constructed during 1995 in upstate New York. It houses several offices for a rural town, as well as a court room and village offices.

The hydronic floor heating system consists of 20 individual circuits of 5/8" PEX tubing served by 5 different manifold locations through the building. The manifold stations are supplied from overhead piping routed above the finish ceiling.

The system is supplied by a single oil-fired boiler with a heating capacity of 120,000 Btuh controlled by outdoor reset control implemented through a 4-way motorized mixing valve. Supply temperature to the floor circuits at design conditions (-15 °F.) is approximately 105 ° F.

The slab is covered by a combination of ceramic tile and glue-down carpet. Floor circuit layout will permit individual room temperature controls to be added in the future. Individual room controls were not included in the original installation to reduce installation cost. However, solar heat gain and occupancy schedules make this a desirable feature for optimal room temperature control.

Figure 2-22 Radiant in 7,000 Sq. Ft. Office Building with -15 °F Design Temperature and 120,000 Btuh Boiler

2

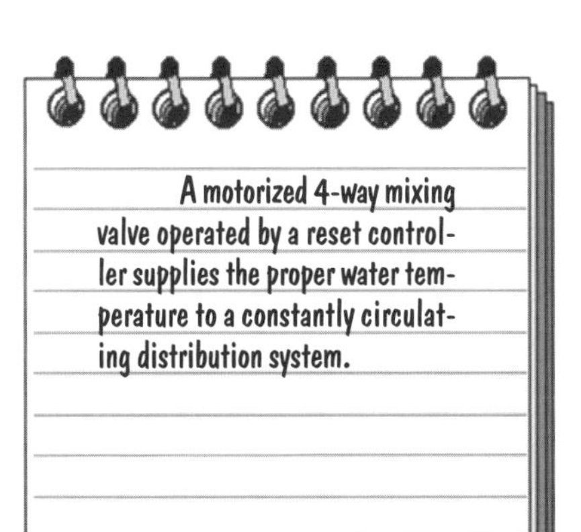

Figure 2-23 Reset Control with 4-Way Mixing Valve

supply temperature sensor

mixed (reduced temperature) water to manifolds

4-way mixing valve, actuator motor

hot boiler water

safety controls

valve motor

controller

outside air temperature sensor

return temperature sensor

Figure 2-24 Piping Schematic for Case Study #6

2

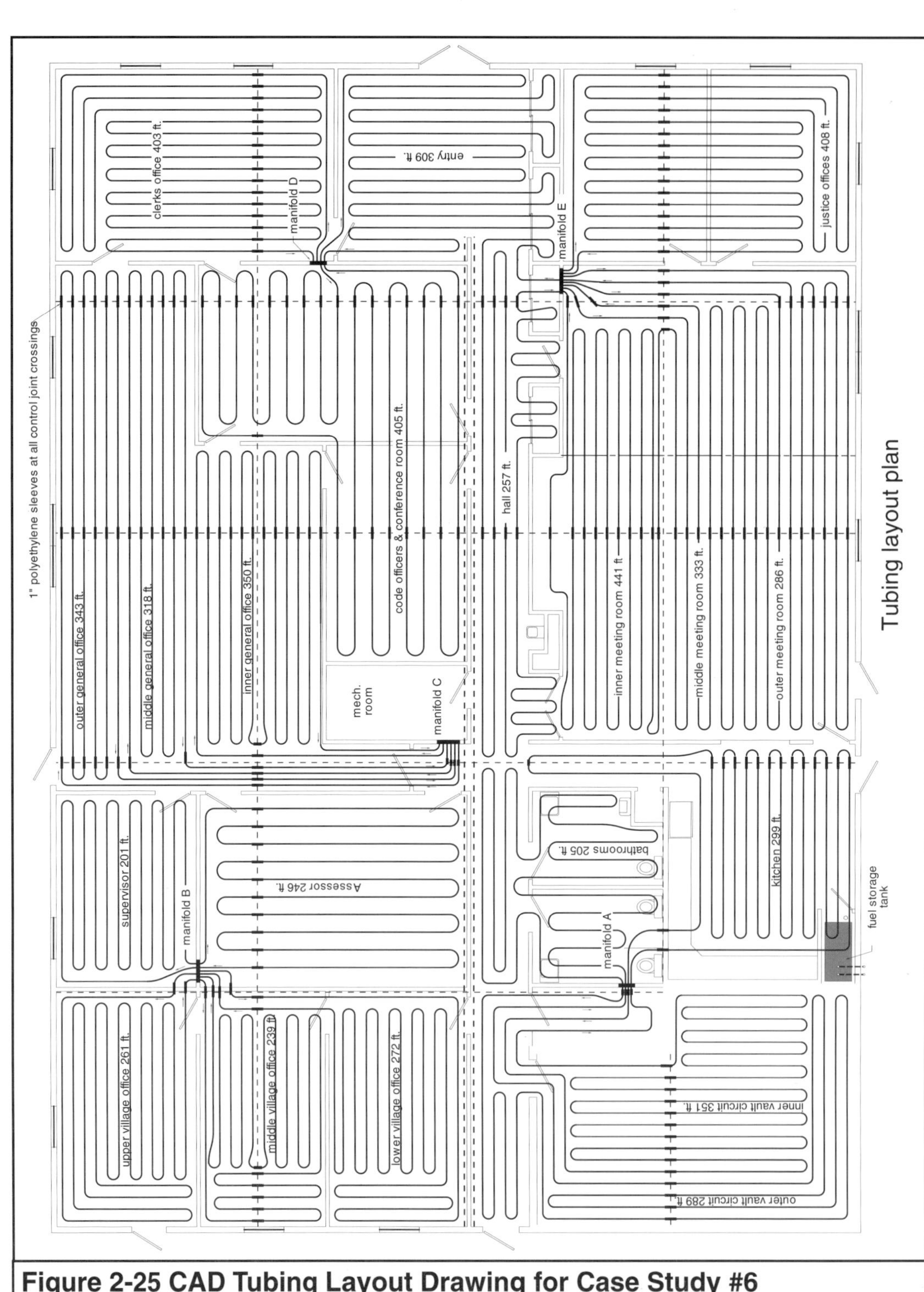

Figure 2-25 CAD Tubing Layout Drawing for Case Study #6

2

CASE STUDY 7
Lumber yard office using hydronic floor heating with injection mixing.

The 6000 square foot building in this case study houses the offices and retail display space for a lumber yard. It was built during 1997 in upstate New York. The owner specifically requested radiant heating based on his favorable experience with electrically-heated gypsum board ceilings in a former house.

The heating system consists of a direct-vent gas-fired boiler supplying a variable-speed pump injection mixing system. Mixed water is routed to two manifold locations. The 4" concrete slab is heated by 16 circuits of 5/8" PEX tubing. Tube spacing varies from 6" to 18" on center. 6" tube spacing was used in the entry foyer and customer side of the counter to encourage rapid drying of tracked-in snow or water. Most of the retail display area uses 12" tube spacing. 18" tube spacing was used within totally interior areas with minimal heating load. The slab is covered with vinyl tile in most public areas, and glued-down commercial grade carpet in the offices. One inch extruded polystyrene insulation is used under all interior areas of the slab, with 2" under the outer 4 foot perimeter floor area and between the edge of the slab and foundation.

The building is controlled as a single zone. The injection mixing systems adjusts water temperature to the floor based on outside temperature (reset control). An indoor temperature sensor provides constant feedback to the injection mixing control, allowing it to adjust for internal heat gains as necessary. Constant water circulation is maintained in the floor circuits. Flow through each floor circuit was balanced using valves built into the return manifolds. The option exists to add more zoning controls to some of the office areas in the future.

Figure 2-26 Lumber Yard Office Building

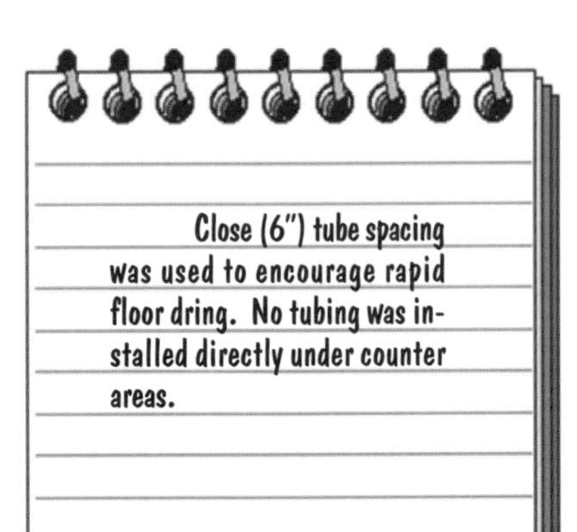

Figure 2-27 No Tubing Installed Under Counter Areas

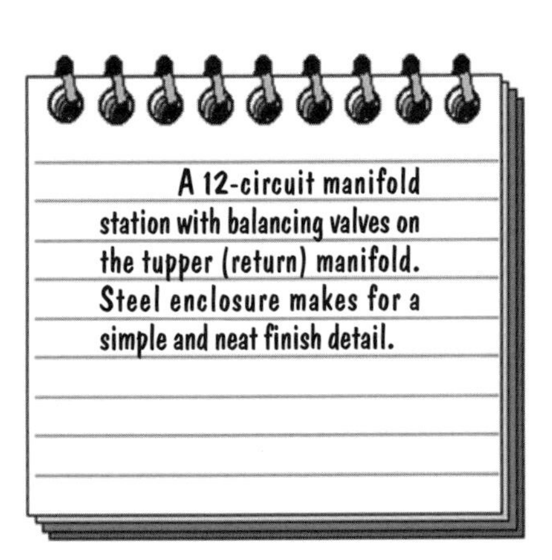

Figure 2-28 Manifold Station with Access Door

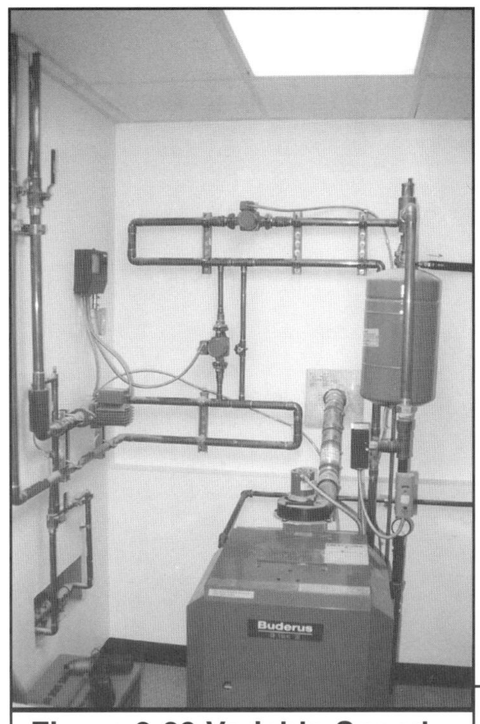

Figure 2-29 Variable-Speed Injection System

Variable-speed injectiom "meters" hot boiler water from the upper primary loop into the low-temperature lower secondary (distribution) loop. Note the closely-spaced tees connecting the injection risers to the primary and secondary loops.

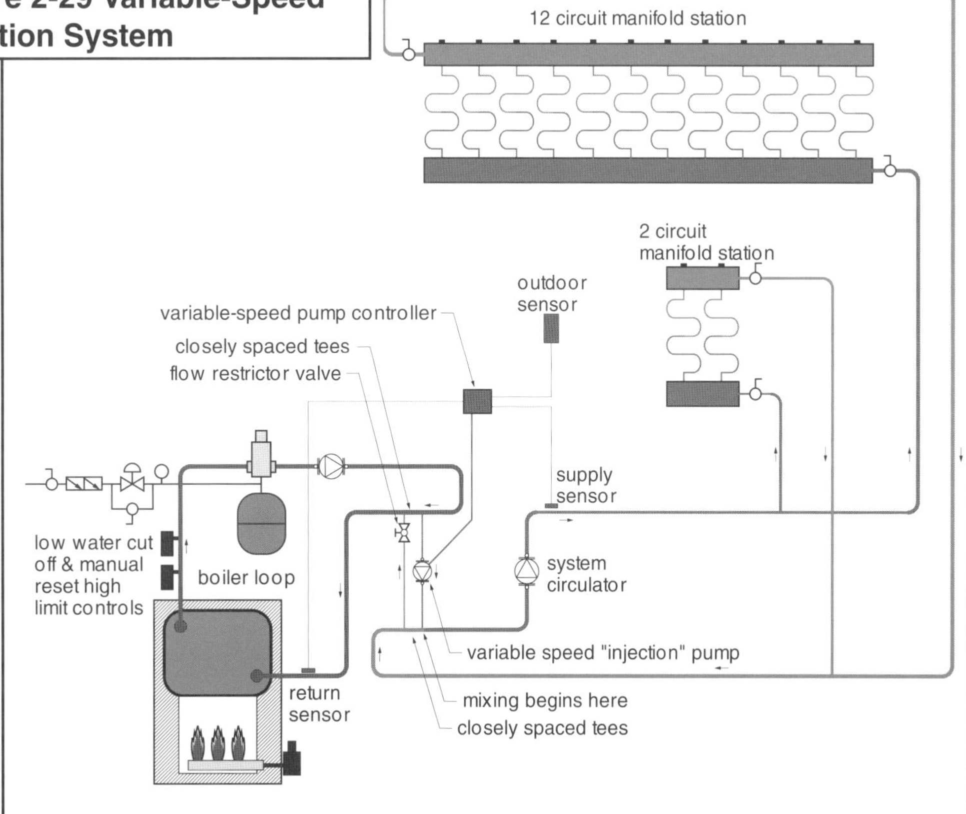

Figure 2-30 Piping Schematic for Case Study #7

2

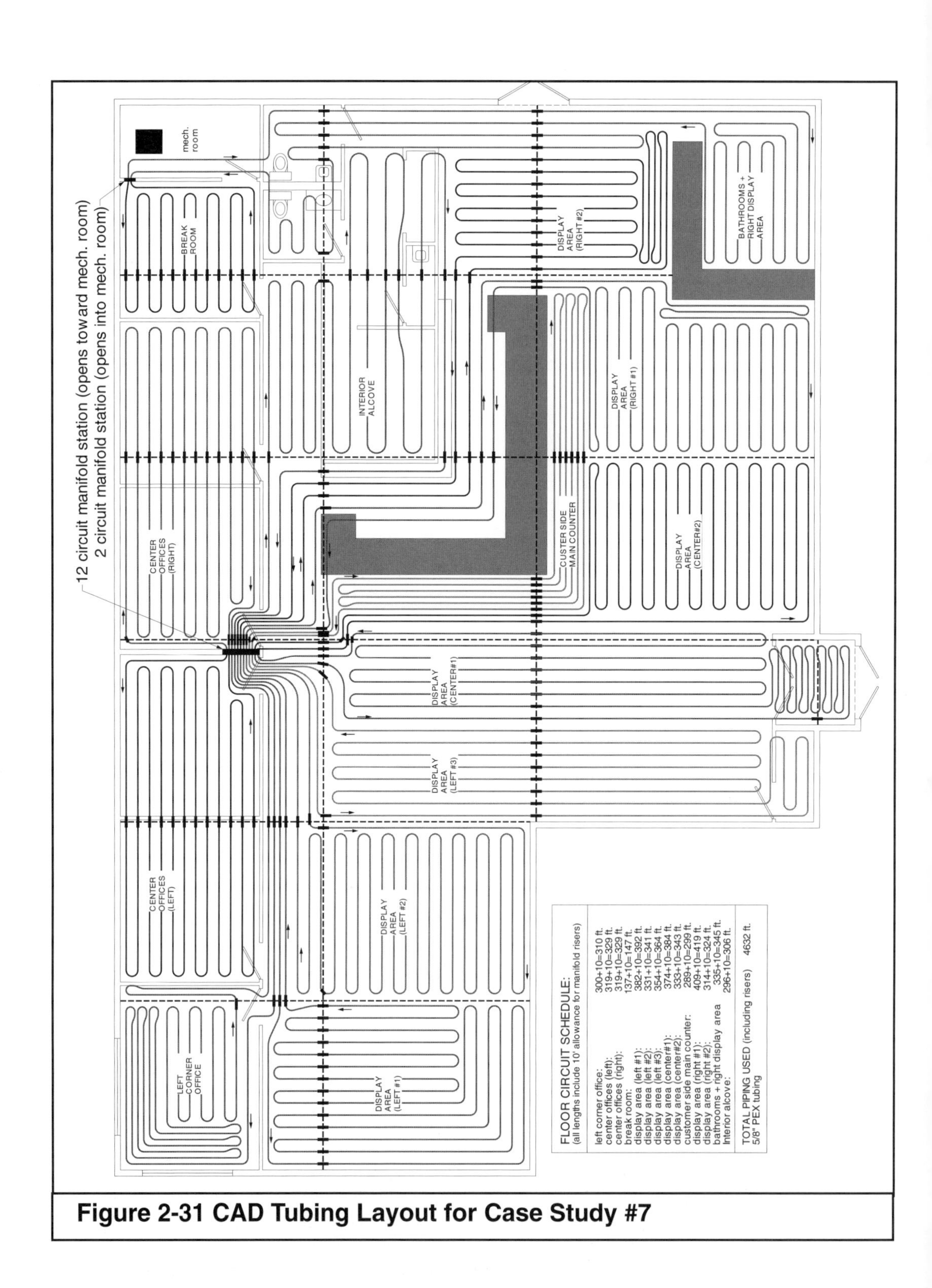

Figure 2-31 CAD Tubing Layout for Case Study #7

2

CASE STUDY 8
Ronald McDonald house using electric radiant floor and ceiling heating.

This Philadelphia, Pennsylvania building serves as a half-way house for parents visiting children undergoing treatment for cancer at local hospitals. Because of the low resistance to illness many of these children have, it was crucial to provide warm floors for them to play on. An electric floor heating system was selected to supplement the building's air-to-air heat pump system.

The radiant system consists of heating cable embedded in concrete slab floors. The floors are covered with carpet and pad. Heat output is approximately 12 watts per square foot (about 41 Btuh per square foot). The total installed capacity of electric floor heating is 80 KW.

The system is supplied through a 3 phase/208 VAC/60 amp disconnect. Low voltage solid-state temperature controller equipped with slab sensors are used in each of three zoned areas of the building. These controls turn on parallel-wired sets of cables through double-pole contactors.

A separate electric snow-melt system is also used for the entrance and sidewalk areas of the building. Heat output from the snow-melt slabs is approximately 55 watts per square foot.

Figure 2-32 Ronald McDonald House in Philadelphia, PA

2

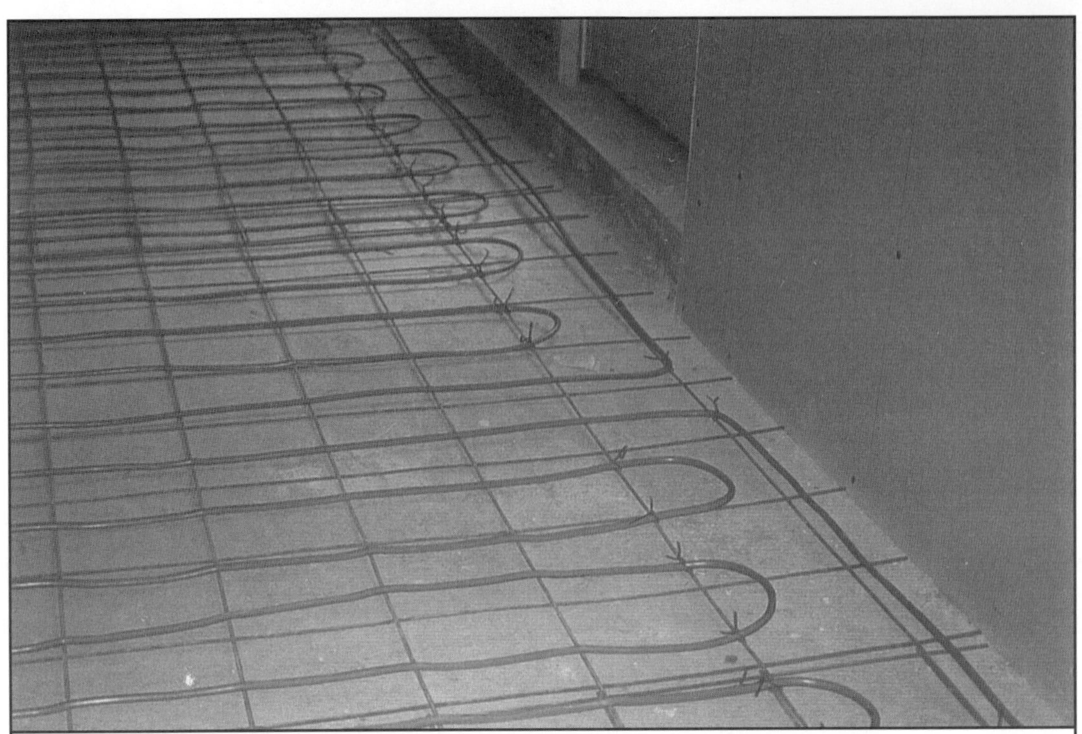

Figure 2-33 Electric Cables Fastened to Welded-Wire Mesh

Figure 2-34 Enclosure for Temperature Controllers and Contactors

24 VAC supplied by transformers powers temperature controllers and double-pole contactors.

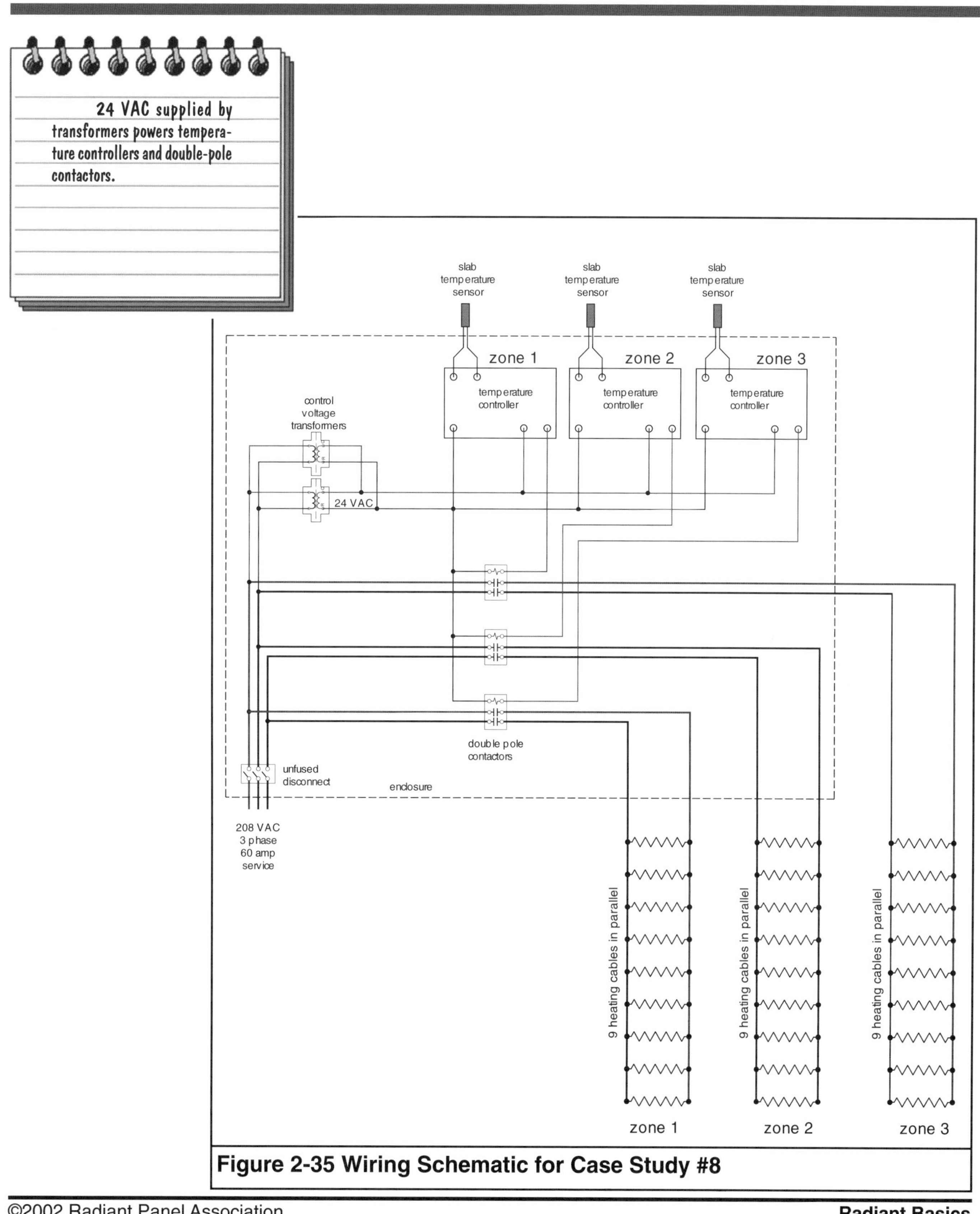

Figure 2-35 Wiring Schematic for Case Study #8

Hydronic Radiant Panel Systems

3•1 Introduction

This section briefly describes many of the common methods for hydronic radiant panel heating. These include site-built assemblies, such as tubing embedded in concrete floors, as well as the installation of manufactured radiant panels.

All heated floors must take finish flooring materials into consideration. The flooring becomes part of the heat delivery system. Floor coverings with high thermal resistance will severely limit the floor's heat output. Although many types of floor coverings are suitable, thick carpet, especially when combined with urethane padding, is not usually recommended for floor heating applications.

For comfort reasons, floor surface temperatures are generally limited to 85 to 87 °F in areas where prolonged foot contact is likely. In other areas where prolonged foot contact is uncommon (hallways, entry foyers, perimeter areas etc.) floor surface temperatures can safely go up to 90 to 92 °F The higher the floor surface temperature the higher the heat output from the floor. An 85 °F floor surface will release heat at about 34 Btuh/sq.ft. into a room at 68 °F If the floor surface averages 92 °F, heat output will increase to about 48 Btuh/sq.ft.

Most hydronic radiant panel systems can be classified as follows:

1. Site-built panels:
   a. Slab-on-grade
   b. Thin-slab systems
   c. Plate systems
   d. Suspended tube systems
   e. Radiant ceiling systems
   f. Radiant wall systems
2. Manufactured panels
   a. Metal wall panels
   b. Metal ceiling panels

Figures 3-1 and 3-2 illustrate the basic configuration of these panels.

Keep in mind that not all hydronic radiant panel systems currently available fit neatly into these categories. In some cases manufacturers have developed specific (and proprietary) materials and installation methods. New products will undoubtedly come onto the market as the industry grows.

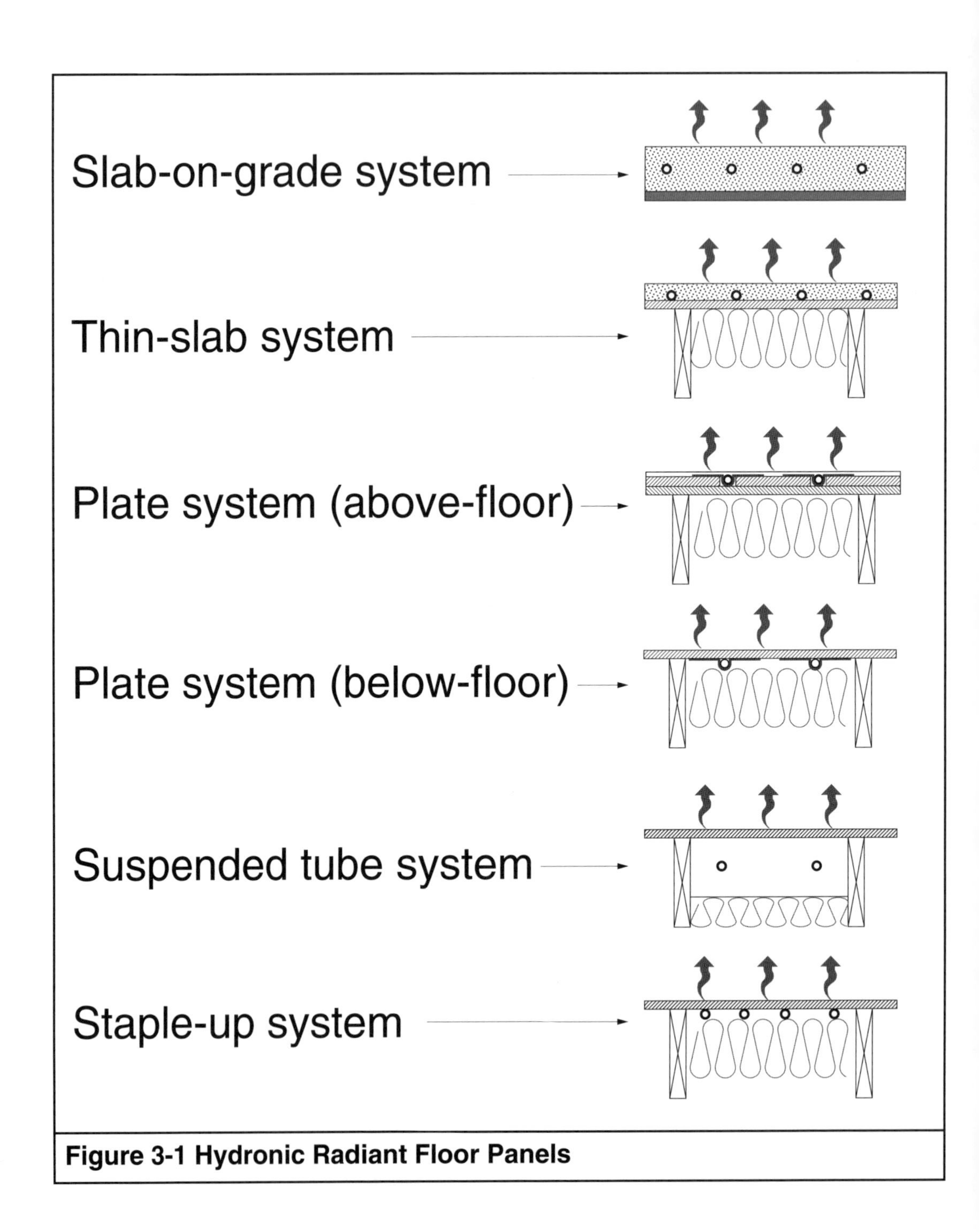

Figure 3-1 Hydronic Radiant Floor Panels

3

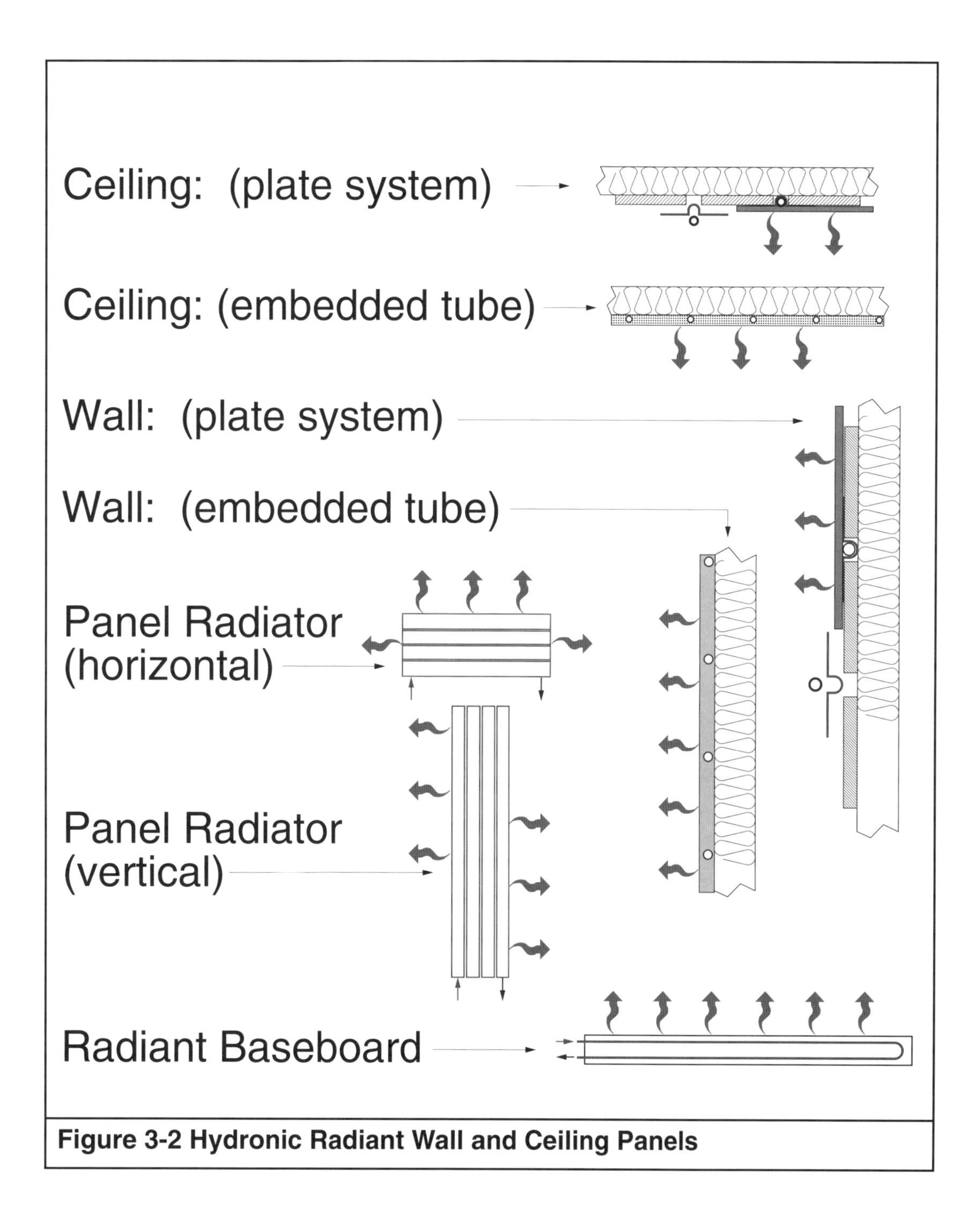

Figure 3-2 Hydronic Radiant Wall and Ceiling Panels

3•2 Slab-on-Grade Systems

The most common type of hydronic radiant panel system is installed in a concrete slab-on-grade floor. It's suitable for a wide range of buildings including homes, commercial/retail buildings, service garages, warehouses, churches, and aircraft hangars to name a few.

The floor system consists of tubing circuits embedded in the slab when it's poured. Warm water from a variety of heat sources is circulated through the tubing circuits. Heat from the tubing disperses into the concrete and eventually into the rooms above. Insulation under the slab and at its edges minimizes losses, directing most of the heat upward.

Heated slab-on-grade floors have operating characteristics that differ substantially from other types of heat. The large thermal mass of such floors makes them slow to respond to wide changes in thermostat settings. This has both good and bad implications. On the plus side, the high thermal mass stabilizes heat output. The mass of the floor stores a considerable amount of heat that, if necessary, can be released much faster than normal. This allows a heated floor to substantially boost its output when cool air from an open door floods in across the floor. It makes a slab-on-grade system well suited for vehicle garages or other uses where there will be frequent openings of exterior doors.

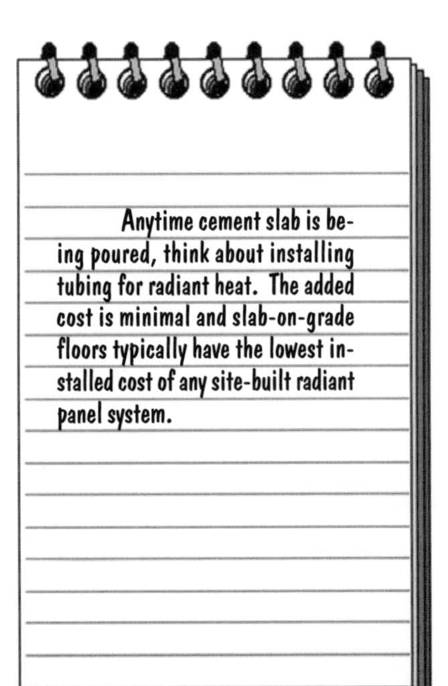

On the negative side, the slow response of a heated slab-on-grade floor significantly limits its use with setback thermostats. Frequent and deep setbacks are difficult to achieve. Setbacks to lower temperatures as well as recovery to normal temperatures must be initiated several hours in advance if satisfactory results are to be attained. Some setback applications can still work well-for example, a church sanctuary that may only be used one day a week and kept at a lower temperature at other times to conserve energy.

The large thermal mass and relatively slow response of a heated slab-on-grade floor can also create overheating in rooms with significant solar heat gain or other internal heat sources. The problem is that once heat is released from the tubing into the slab it is no longer "stoppable" by the system controls. Its output will be stopped only when the solar heat gain has driven room temperature up to the surface temperature of the floor, which is well above normal comfort temperature. In general, passive solar buildings are better served by heating systems with low thermal mass and thus faster temperature response.

3

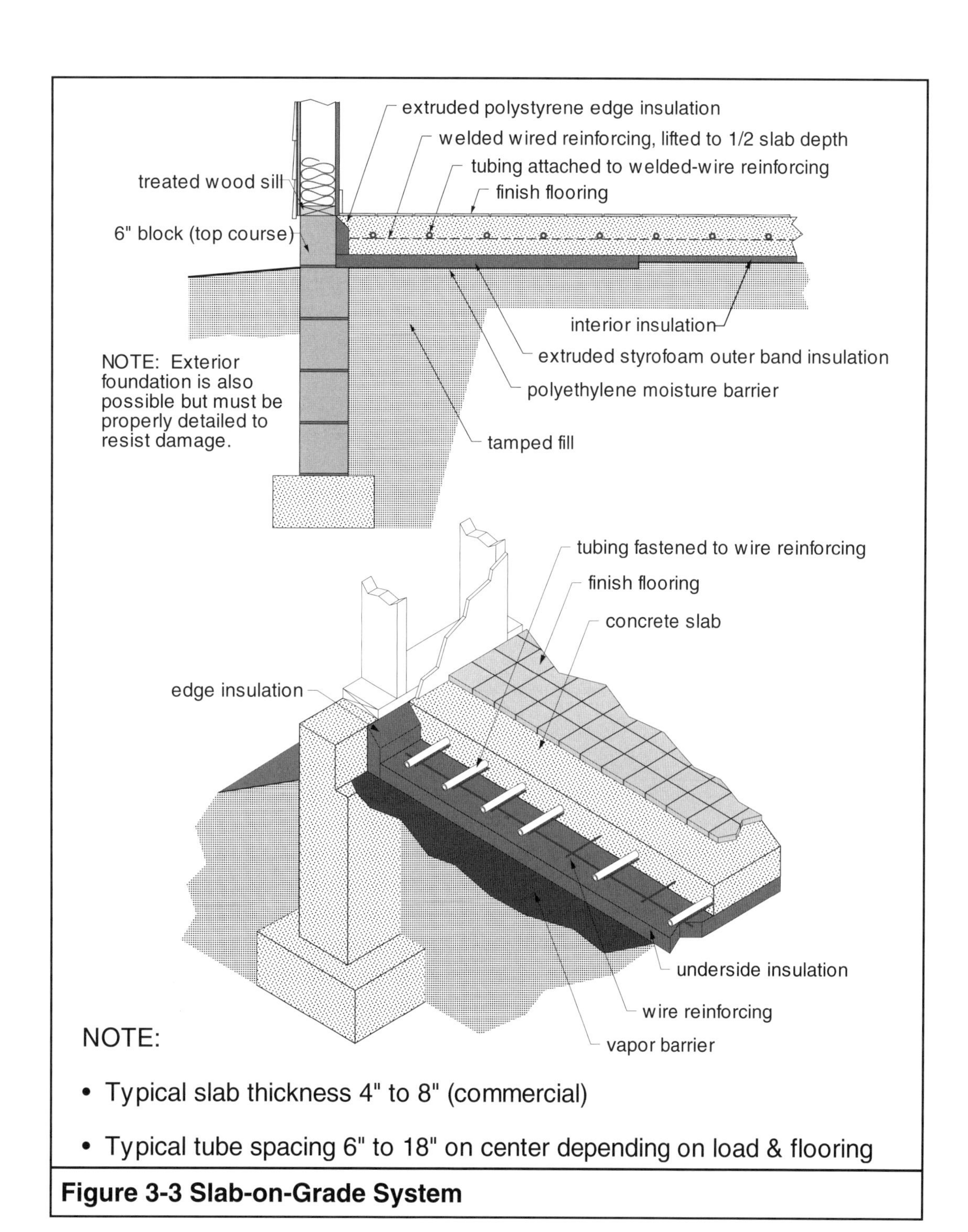

NOTE:

- Typical slab thickness 4" to 8" (commercial)
- Typical tube spacing 6" to 18" on center depending on load & flooring

Figure 3-3 Slab-on-Grade System

3

3•3 Thin-Slab Systems

Obviously not all buildings have slab-on-grade floors. However, several methods are available for heating wood-framed floors. One method is known as a thin-slab system. The same type of tubing used in a slab-on-grade system is fastened to the top of the subfloor and then covered with a thin poured slab. Such slabs are typically 1.25" to 2" thick and consist of either poured gypsum underlayment or a special concrete mix. The slab material encases the tubing and allows the heat to spread laterally away from it.

Depending on the slab material and thickness, thin-slabs can add from 10 to 20 pounds per square foot (PSF) to the dead loading of the floor structure. This added weight must be accounted for in the structural design of the floor. It is suggested that floor framing be designed to have a minimum live load deflection of 1/480th of the clear span of the floor joists, with a preferable maximum deflection of 1/600th of the clear span. This criteria stiffens the floor and helps prevent excessive deflection or floor vibration.

In addition to weight, other architectural factors must be considered when using a thin-slab system. Since the floor level will be raised 1.25" to 2", the heights of rough openings for doors and windows should be adjusted if they are framed prior to pouring the thin-slab. The heights of stair risers will also have to be adjusted to accommodate the new floor height. Base cabinets for kitchens and bathrooms in which thin-slabs will be installed should be "furred up" to the thickness of the slab. This furring creates a dam that prevents the thin-slab material from filling areas under cabinets where heat output is ineffective. These dams also prevent the thin-slab material from pouring over the edges of stairwells or other similar openings in the floor. Closet flanges for toilets also need to be set higher to accommodate the added slab thickness.

Because thin-slabs affect a number of architectural factors in the building, it's always best to discuss their possible use early in the building's planning. All of the factors mentioned above can be properly accommodated during building design if the designer is aware of them. If you're considering retrofitting a thin-slab system to an existing building, these same factors must be addressed, especially the added weight of the thin-slab. If in doubt, have a competent structural engineer verify the floor's load carrying ability prior to proceeding further.

Figure 3-4 shows a typical thin-slab constructed of poured gypsum underlayment. Manufacturers of poured gypsum underlayments publish specific information about the materials and methods used for attaching finish flooring to the completed slab. These guidelines specify when sealing the floor is necessary, and include adhesive recommendations for various floor covering applications. It is strongly recommended that such procedures be obtained and followed.

Concrete Thin-slabs:

Thin slabs can also be constructed of portland cement based concrete as shown in Figure 3-5. The same architectural considerations that apply to gypsum thin-slabs must also be observed with concrete thin slabs.

One concrete mix that has been used for heated thin-slab applications is given

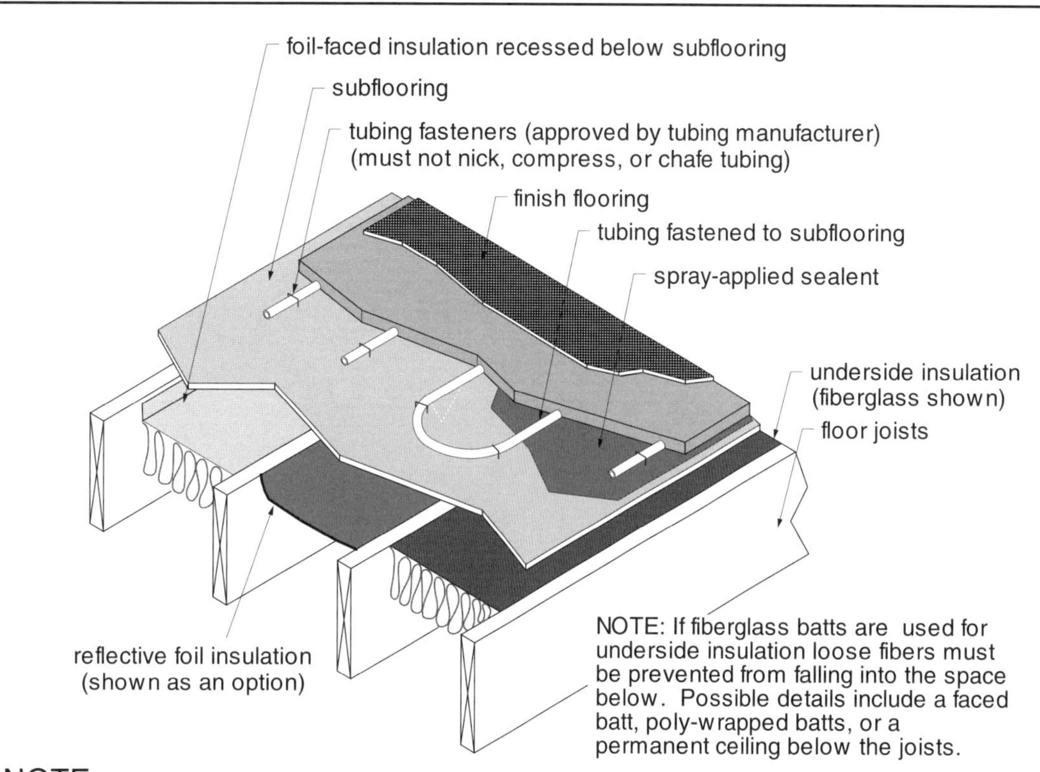

NOTE:

- Typical slab thickness: 1.25" to 1.5"
- Typical loading created by slab:
  14.5 pounds / sq. ft. added dead loading at 1.5" thickness
- Typical tube spacings: 4" to 12"
- Minimum underside insulation is R-11 above heated space, R-19 above partially heated space, R-30 above unheated crawl space.
- Maximum floor deflection (under full live loading):

Figure 3-4 Gypsum Thin-Slab

in Figure 3-6. Admixtures such as a superplasticizer, water reducing agent, and fiberglass thread reinforcing give the concrete the "flowability" to encase the tubing, as well as low shrinkage characteristics that minimize cracking. This formulation of concrete weighs about 140 pounds per cubic foot. It adds about 18 pounds per square foot (PSF) to the dead loading of the floor at a thickness of 1.5 inches.

3

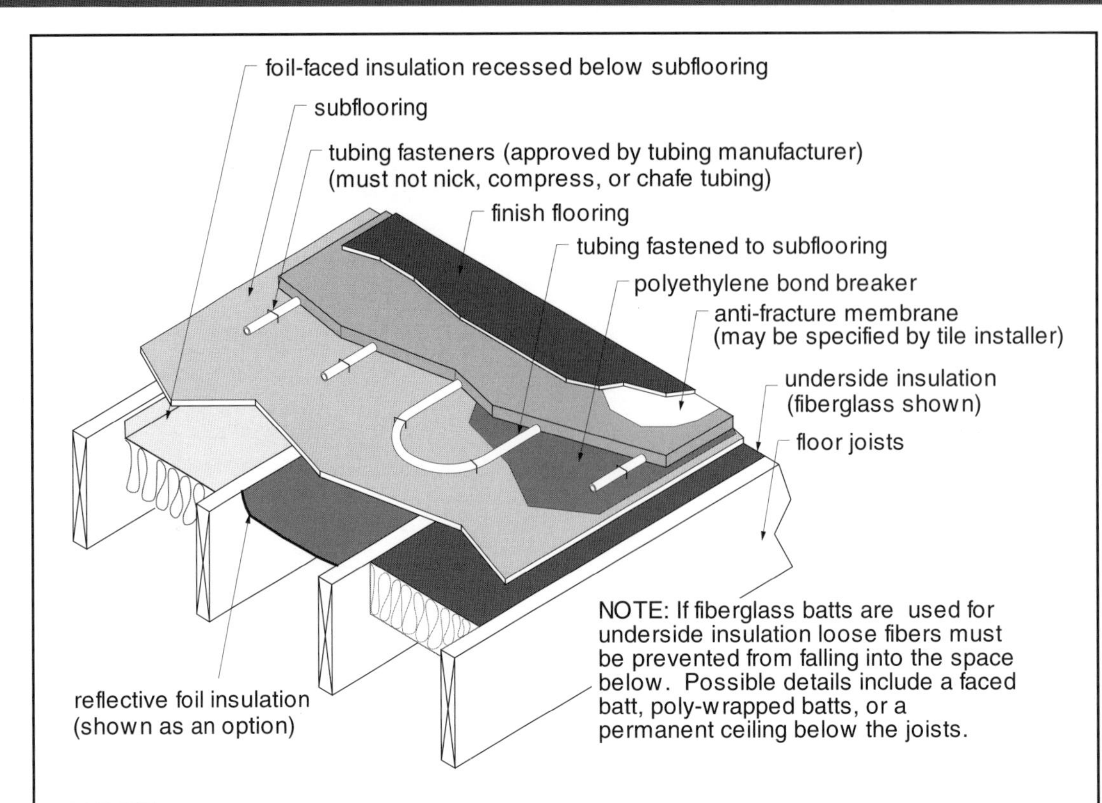

NOTE:

- Typical slab thickness: 1.5" to 2"
- Typical loading created by slab:
  18 pounds / sq. ft. added dead loading at 1.5" thickness
- Typical tube spacings: 4" to 12"
- Minimum underside insulation is R-11 above heated space, R-19 above partially heated space, R-30 above unheated crawl space.
- Maximum floor deflection (under full live loading):

Figure 3-5 Concrete Thin-Slab

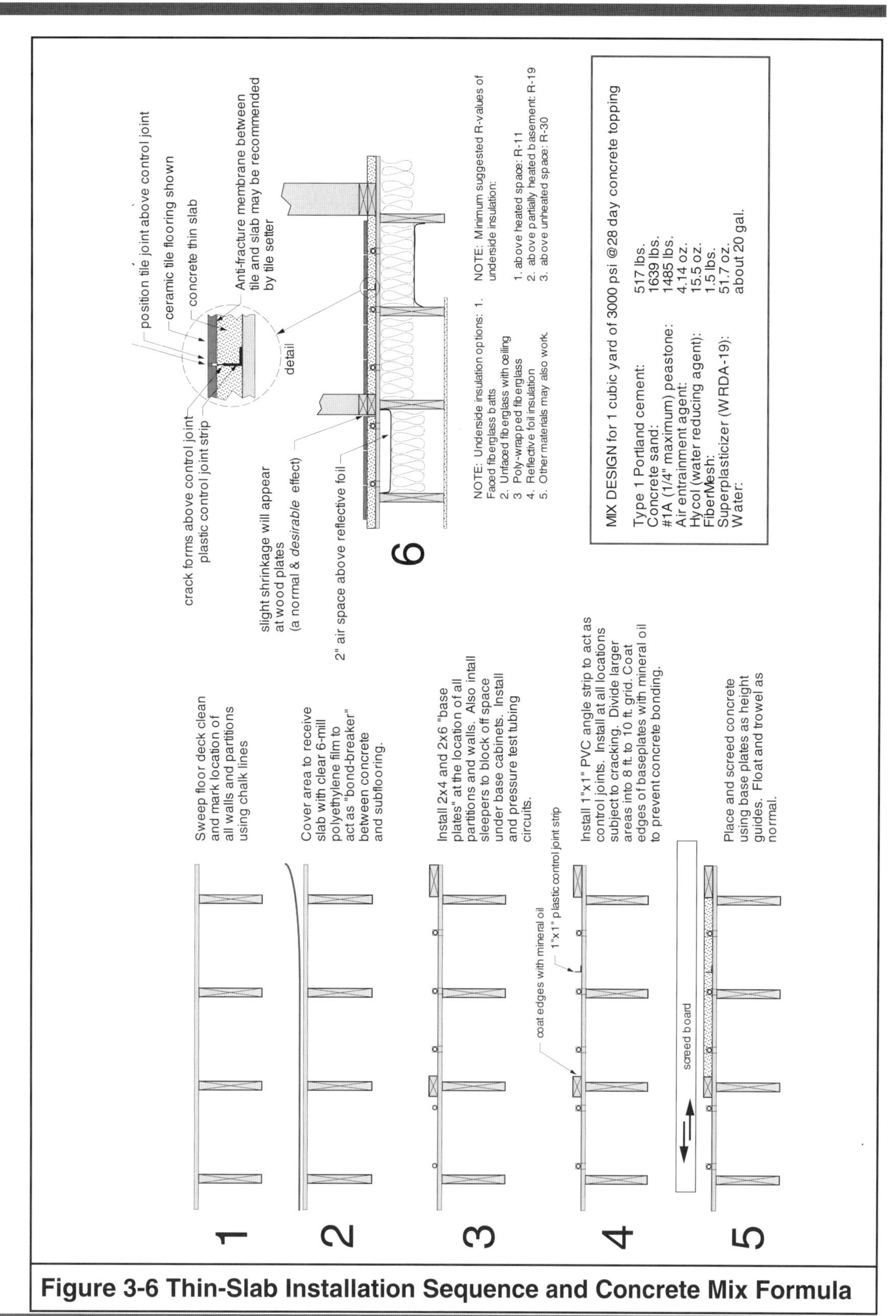

Figure 3-6 Thin-Slab Installation Sequence and Concrete Mix Formula

3•4 Plate Systems and Sandwich Method

Many opportunities for installing hydronic radiant floor heating will not allow the added thickness and weight of a thin-slab system. In such cases an aluminum "heat transfer plate" that partially surrounds the tubing can be used to conduct the heat laterally away from the tube and spread it across the floor. The high conductivity of aluminum in comparison to poured thin-slab materials allows even a thin heat transfer plate to adequately spread the heat under the flooring. Such a "plate system" will add very little weight to the floor and lends itself to both new and retrofit applications.

Above-floor Plate System:

Heat transfer plates can be installed either above or below a wood-framed floor. In an above-floor installation the subfloor is prepared by installing spaced wooden "sleepers" to create cavities into which the tubing (and a portion of the heat transfer plate) will be installed. The sleepers are typically made of furring-grade wood or plywood, and should be glued and mechanically fastened to the subfloor to prevent squeaks. The width of the sleepers should be adequate to support the full width of the heat transfer plates. Semicircular grooves are created at the ends of straight tubing runs to allow a serpentine circuit pattern to be formed.

The heat transfer plates are placed in the grooves between the sleepers and fastened (on one side only) with a staple gun. A minimum of 1/4" should be left between the edges of adjacent plates to allow for expansion. The tubing is then laid in place and snaps down into the groove in the heat transfer plate.

After the circuits are pressure tested, the floor is usually covered with a thin (1/4" or 3/8") plywood cover sheet. This creates a smooth substrate for the subsequent installation of ceramic tile, carpet, or vinyl flooring. When solid hardwood flooring will be nailed down perpendicular to the straight tubing runs, the cover plate is usually unnecessary. See Figure 3-7 for a detail of a typical above-floor plate-type installation.

Above-floor installations allow the tubing to be routed in the most advantageous direction without concern for the direction of the floor framing. The 1/2" to 1" increase in floor thickness is usually easy to accommodate without major architectural changes. Areas of the floor that are covered with built-in cabinets are not equipped with tubing and plates since heat output would be ineffective. Any nailing, sawing, or drilling into the floor must obviously be done with care so as not to damage the tubing.

Above-floor System Without Plates (Sandwich Method):

In an above-floor "sandwich method" installation, the subfloor is prepared similarly to the floor plate system by installing spaced wooden "sleepers" to create cavities into which the tubing will be installed. The sleepers are typically made of furring-grade wood or plywood, and should be glued and mechanically fastened to the subfloor to prevent squeaks. The width of the sleepers should be adequate to allow attachment of flooring goods on top. Semicircular grooves are created at the ends of straight tubing runs to allow a serpentine circuit pattern to be formed.

After the circuits are pressure tested, the floor is usually covered with a thin (1/4" or 3/8") plywood cover sheet. This cre-

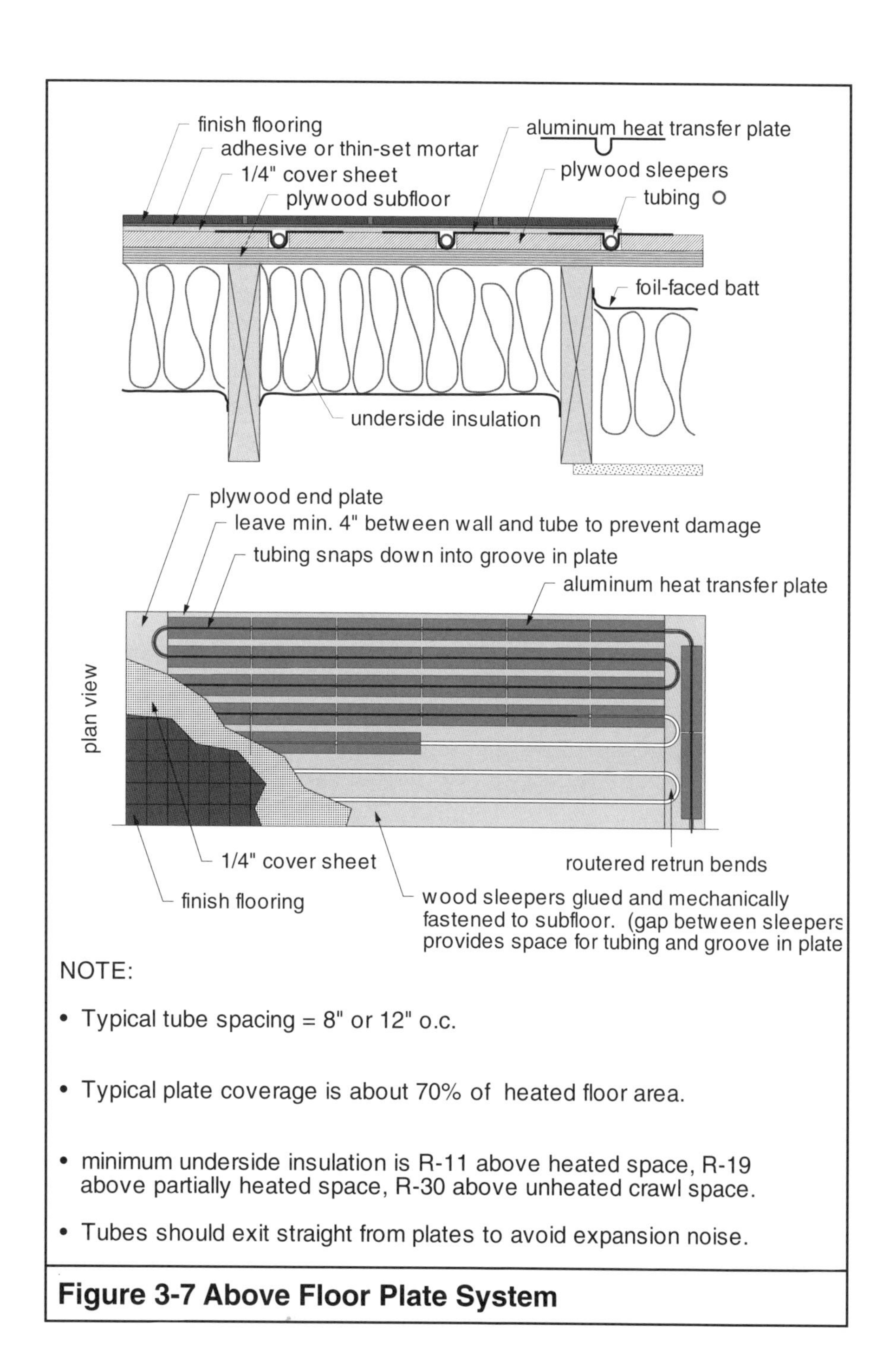

NOTE:

- Typical tube spacing = 8" or 12" o.c.
- Typical plate coverage is about 70% of heated floor area.
- minimum underside insulation is R-11 above heated space, R-19 above partially heated space, R-30 above unheated crawl space.
- Tubes should exit straight from plates to avoid expansion noise.

Figure 3-7 Above Floor Plate System

ates a smooth substrate for the subsequent installation of ceramic tile, carpet, or vinyl flooring. When solid hardwood flooring will be nailed down perpendicular to the straight tubing runs, the cover plate is usually unnecessary. The performance of a sandwich system without plates is more limited in output than with plates and has more temperature banding. However the additional cost of the plates is eliminated.

Below-floor Plate System:

Aluminum heat transfer plates can also be used to support tubing up against the underside of a wood subfloor. This approach eliminates the need for sleepers or cover plates as described for above-floor plate systems. The tubing and plates are usually installed parallel to the floor framing. At the ends of a joist cavity, the tubing is routed through oversized holes and into the next cavity. Special installation methods have been developed for such applications by tubing manufacturers. They allow the tubing to be pulled through the floor framing without kinking or binding. A typical below-floor plate system is shown in Figure 3-8.

Once the tubing has been pulled into the joist cavities, the heat transfer plates are used to clamp it tightly up against the underside of the subfloor. The plates are then stapled in place.

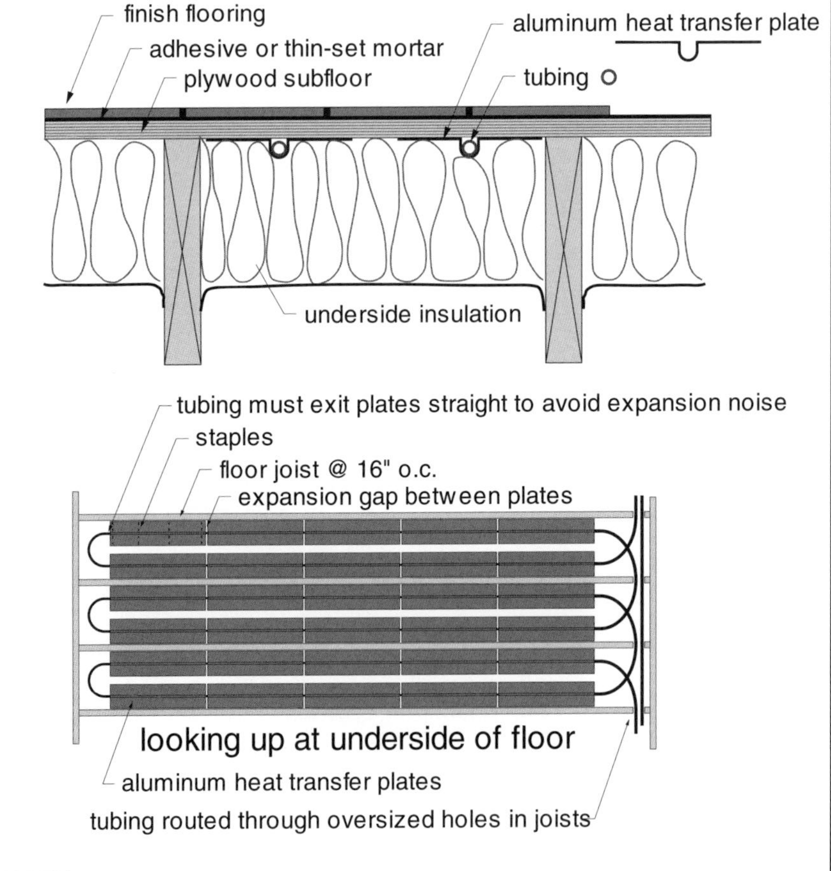

Figure 3-8 Below Floor Plate System

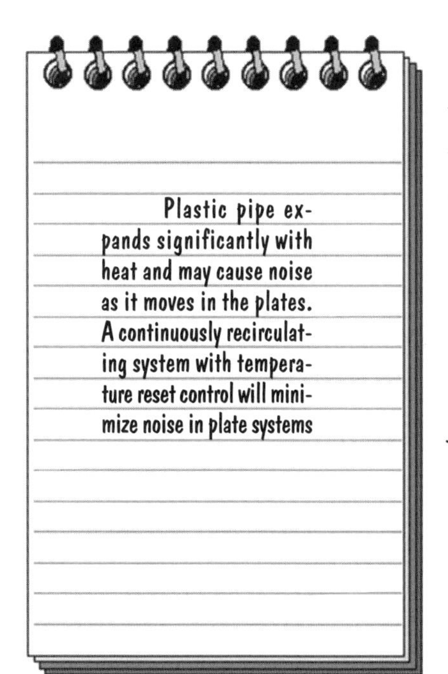

Common tube spacings are 8" o.c. (when two tubes will be placed symmetrically between floor joists, 16" o.c.; occasionally the tubing and plates are also placed at 12" o.c.) Spacings must be selected that allow the tubing to be installed from one joist cavity to the next without interference from the joists themselves.

When evaluating the suitability of a below-floor plate system one should always carefully examine the underfloor. In some cases nail points will be protruding from the underside of the subfloor. Although it is possible to cut or grind them off so the tubing and plates can lie flat, this is very time consuming. In other cases, existing plumbing, wiring, or even oddly installed framing will make the installation more difficult than anticipated.

3•5 Suspended Tube Systems

Another method of installing hydronic tubing in floor cavities is to simply suspend the tubing within the joist cavity as shown in Figure 3-9. Since no slab or heat transfer plate is used, all heat from the tubing is transferred to the underside of the subflooring by a combination of convective air currents within the joist cavity and direct radiant heat exchange. Because of the limited surface area of the tubing, relatively high water temperatures are usually required to generate reasonable heat outputs. In many cases this requires a heat source capable of producing as much as 180 °F water. It also requires tubing capable of sustained operation at such temperatures. To limit downward heat losses it is critical to insulate the bottom of the joist cavities.

Suspended tube systems offer several potential benefits:

- Since the tubing doesn't contact the bottom of the subfloor, it doesn't create high temperature "stripes" that could potentially damage the flooring.
- They eliminate the need and expense of heat transfer plates (and in some cases related materials such as sleepers and cover sheets)
- Nail points protruding from the underside of the subflooring will not contact the tubing and thus are not a concern.
- In some cases high temperature water directly from a boiler can be routed through the tubing without need for a mixing device. This simplifies controls and reduces installation costs.

Depending on insulation details, suspended tube systems may heat the floor joists to temperatures in the range of 110 to 120 °F This will lower the moisture con-

tent of the joists and cause some shrinkage, especially if the moisture level of the joists is initially high. It is recommended that such systems be put into operation several days prior to installing any finish flooring that could be damaged by such shrinkage. A moisture meter should be used to measure the moisture content of the joists, and finish flooring should be postponed until the moisture level is stable. Because of relatively high operating temperatures, and to prevent unbalanced shrinkage, the tubing used in a suspended tube application should not be directly fastened to the joists.

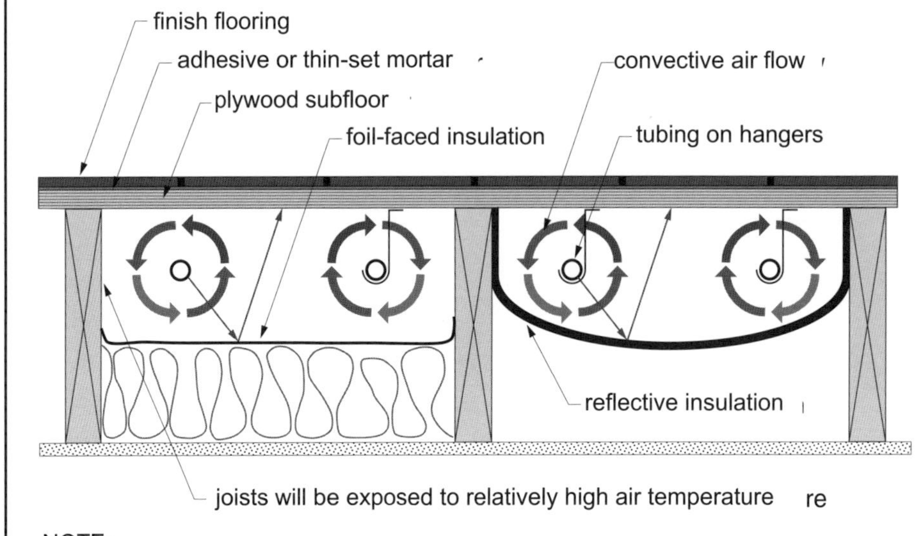

NOTE:
- Tubes operate at high water temperature (160 F to 180 F)
- Heat transfers to underside of subfloor by convection and radiation
- Practical heat output limit is about 20 Btuh/sq. ft. (with bare tubes)
- Some systems use metal fins on tubes to increase convection
- Air in joist space could reach temperatures in excess of 125 F. If this air is in direct contact with joist, some shrinkage is likely. Verify that moisture content of joists is stable prior to installing flooring.
- Minimum underside insulation is R-11 above heated space, R-19 above partially heated space, R-30 above unheated crawl space.
- Be sure there is no air infiltration into joist space
- Plumbing traps should be below heated air space to limit evaporation

Figure 3-9 Suspended Tube System

3•6 Staple-Up Systems

In certain situations tubing can be stapled directly to the bottom of the subfloor without the use of heat transfer plates. The successful use of this approach depends on several factors, including the required heat output of the floor, the thermal resistance of the subfloor and finish floor, and the ability of the finish flooring to withstand variations in temperature.

Because the tubing is installed without a slab or heat transfer plates, lateral heat flow away from the tube is much more limited. This can be partially compensated for by increasing the circuit's water temperature. However, high water temperatures can create temperature "striping" on the floor surface. This describes the situation where the floor surface is quite warm di-

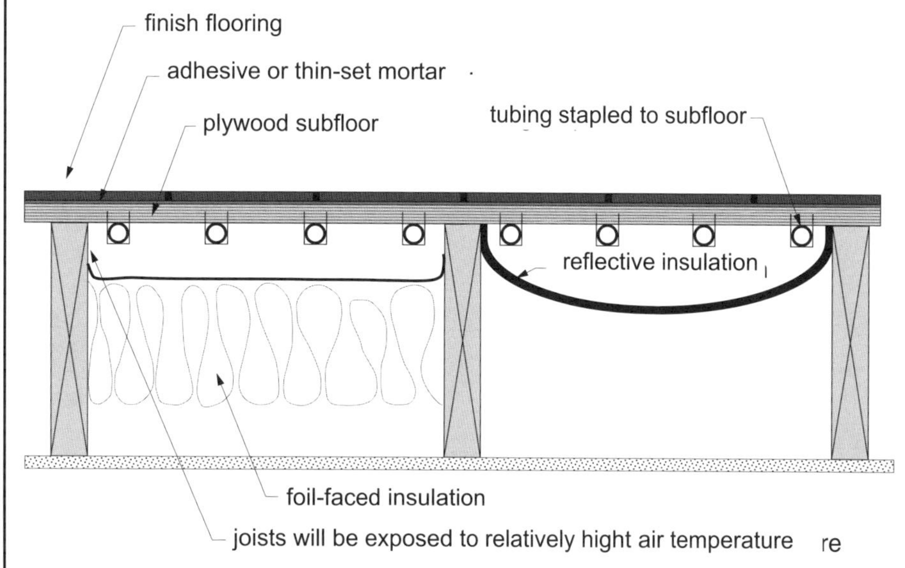

NOTE:

- Tubes typically operate at high water temperature (140+ F)
- Heat transfers to underside of subfloor by conduction, convection and radiation
- Closer tube spacing is common to help spread heat over floor
- Air in joist space could reach temperatures in excess of 125 F. If this air is in direct contact with joist, some shrinkage is likely. Verify that moisture content of joists is stable prior to installing flooring.
- Minimum underside insulation is R-11 above heated space, R-19 above partially heated space, R-30 above unheated crawl space.
- Be sure there is no air infiltration into space above insulation
- Plumbing traps should be below heated air space to limit evaporation

Figure 3-10 Staple-Up System

rectly above the tube, but not very warm a few inches on either side of the tube. To correct for this, closer tube spacing is often recommended. Tube spacings of 4" to 8" are common in staple-up systems. Figure 3-10 shows a typical staple-up installation.

The tubing would be pulled into the joist cavities as previously described for the below-floor plate system. It is then stapled directly to the underside of the subfloor with pneumatic staplers equipped with a special attachment to prevent damage to the tubing.

Staple-up systems tend to be more practical in rooms requiring low to moderate heat output from the floor. If you're considering use of a staple-up system, be sure the finish flooring can withstand significant lateral temperature variations. It's also advisable to operate the system for a few days and verify that the moisture content of the subfloor is stable prior to installing any finish flooring.

3•7 Engineered Subfloor and Board Systems

Modular installation of low mass radiant heating systems may be done in several ways. Modular systems incorporating boards and metal, for improved heat transfer, are available and easy to install. These systems are low in profile and accelerate rapidly. They allow a contractor to sequence a job without waiting for another trade such as cement. These systems are of two types: engineered subfloor, where the subfloor is designed to have a path for the tubing; and board systems, where modular boards are attached on top of the subfloor by gluing, screwing, nailing or stapling. Tube spacing in these systems is fixed to what is provided by the manufacturer which somewhat limits how tubing may be laid out.

These systems have several potential benefits:

- They accelerate more rapidly than thermal mass systems and are more suitable for control with setbacks.
- In climates that may include a heating and a cooling load in the same day, the lack of thermal mass is an advantage
- Contractors can have closer control over job sequencing.
- Since the tubing is located just under the floor coverings, the system is transferring heat through low R-values and performs well.

Figure 3-11 shows an engineered subfloor. Premanufactured 1-1/8" thick panels have grooves for tubing and an aluminum sheet bonded to the board. In this case, the premanufactured panels serve

Figure 3-11 Engineered Subfloor

both as the structural subfloor and as the channel into which the tubing is in stalled. The aluminum sheet makes the system accelerate rapidly and spreads out the heat. Tubing is normally installed 12" on center in grooves.

This product must be specified early in a project since it is the subfloor and is normally installed during the initial framing. This product also usually requires that tubing be installed early in a project. When this occurs the tubing must be protected from damage during the remainder of construction. The boards must be carefully aligned and laid out. Figure 3-12 shows the installation of an engineered subfloor system. A CAD layout of the placement panels is recommended (Figure 3-13). Panels come in 4' x 8' and can be cut with a saw. Non-standard thickness of the subfloor must be accounted for in building elevations and construction. Some cut panels need blocking for support as recommended by manufacturer. Since this board is the subfloor, it must be installed throughout the entire radiant area. Some floor coverings may require a layer of backer board on top of radiant board system.

Figure 3-12 Installation of an Engineered Subfloor System

3

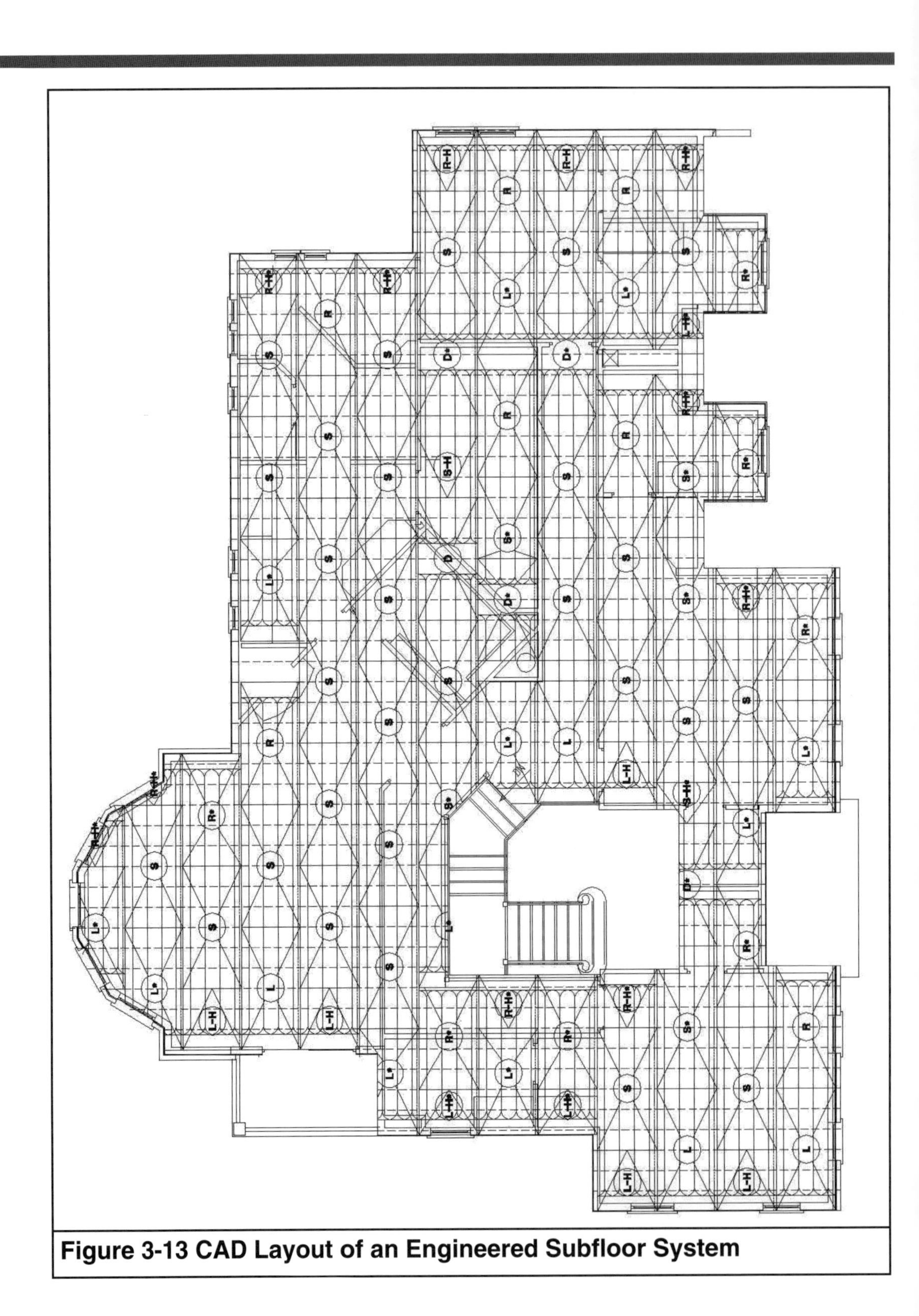

Figure 3-13 CAD Layout of an Engineered Subfloor System

Figure 3-14 shows an installation of a modular board system on top of the subfloor.

Figure 3-14 Modular Board System

Two modular board variations are currently available. One board has metal on the bottom and the other on the top. Both serve to spread the heat laterally. Normally they are glued and screwed or stapled to the top of a wooden subfloor. Under some conditions they may be attached on top of existing slabs. Modular systems include boards with grooves cut in straight and curved patterns that are assembled to make a channel for pipe. Different products use different pipe sizes.

The boards must be carefully aligned and laid out. Spacing of pipe, board size and thickness varies from product to product. Typically boards are 1/2" or 5/8" thick, making them useful in retrofit applications as well as new construction. They can be installed quite late in job sequence or just after sheetrock to provide job heat. A CAD layout is helpful. Thickness of boards needs to be accounted for in cabinet toe kicks, stair rise and runs. Some floor coverings may require a layer of backer board on top of radiant board system.

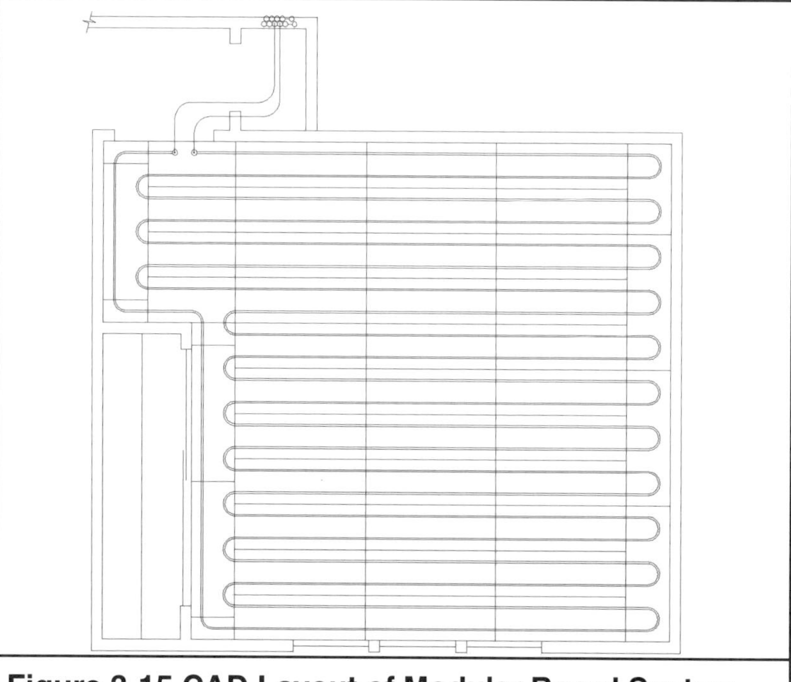

Figure 3-15 CAD Layout of Modular Board System

3•8 Ceiling Heating

Like heated floors, hydronically heated ceilings have also been used in the USA for several decades. Some systems installed during the 1940's and 1950's used copper or steel pipe embedded in "wet plaster" ceilings. Others used suspended high output metal panels with copper tubing mechanically attached to their topside. Many of these systems are still functioning today.

Hydronically heated ceiling systems can be site built as is typical in residential applications or they may utilize commercially available modular panels. The modular panels are often placed in a suspended ceiling grid in commercial construction. In large commercial projects that have proper air movement and controlled dehumidification, it is possible to chill the panels for cooling in the warmer months. In this case the panels are usually sized for cooling since this function often takes more surface area. These systems are covered in Section 3•10.

Heated ceilings have several unique characteristics that distinguish them from heated floors.

- **Heated ceilings can operate at higher surface temperatures than heated floors:** Because occupants are not in direct contact with them, ceilings can safely operate at higher temperatures without causing discomfort. This allows significantly higher heat outputs per square foot.. For example, a radiant ceiling operating at an average surface temperature of 102 °F° will have a heat output of about 55 Btuh/sq. ft. in a room with an air temperature of 68 °F. This is almost 60% more heat output than a floor with a surface temperature limit of 85 °F can yield in the same room.
- **Heated ceilings are not affected by floor coverings and furniture.** Because they are not covered with flooring materials or furniture, heated ceilings provide a large unobstructed panel from which radiant heat can be delivered to the room below. During the life of many buildings there's the possibility that floor coverings and furniture placement will change. This could affect the output of a heated floor, but will not affect the output of a radiant ceiling.
- **Most heated ceilings respond faster than heated floors.** Because they have low thermal mass, heated ceilings warm up rapidly when heat is called for by a room thermostat. This characteristic also allows them to stop emitting heat faster than slab-type floor systems when room comfort is achieved. For example, a heated ceiling in a room that suddenly experiences strong solar heat gains could stop releasing heat quickly and thus limit any temperature overshoot. Low mass ceiling heating also allows deeper and more frequent temperature setbacks in areas that don't require normal comfort temperatures at all times.
- **Radiant ceilings will warm the floor as well as objects in the room below.** Remember that radiant heat travels equally well in all directions. Just as light given off by a ceiling lamp shines down on the floor, so does radiant heat

(infrared light). There can, however, be some “shadow effects” from radiant ceiling heat. A large horizontal surface (such as a desk top) can block direct radiant heat transfer to the space underneath it. Although this effect is generally not a serious problem in modern buildings with good insulation, the designer should be aware of it. Under high heating load conditions this shadow effect may not provide

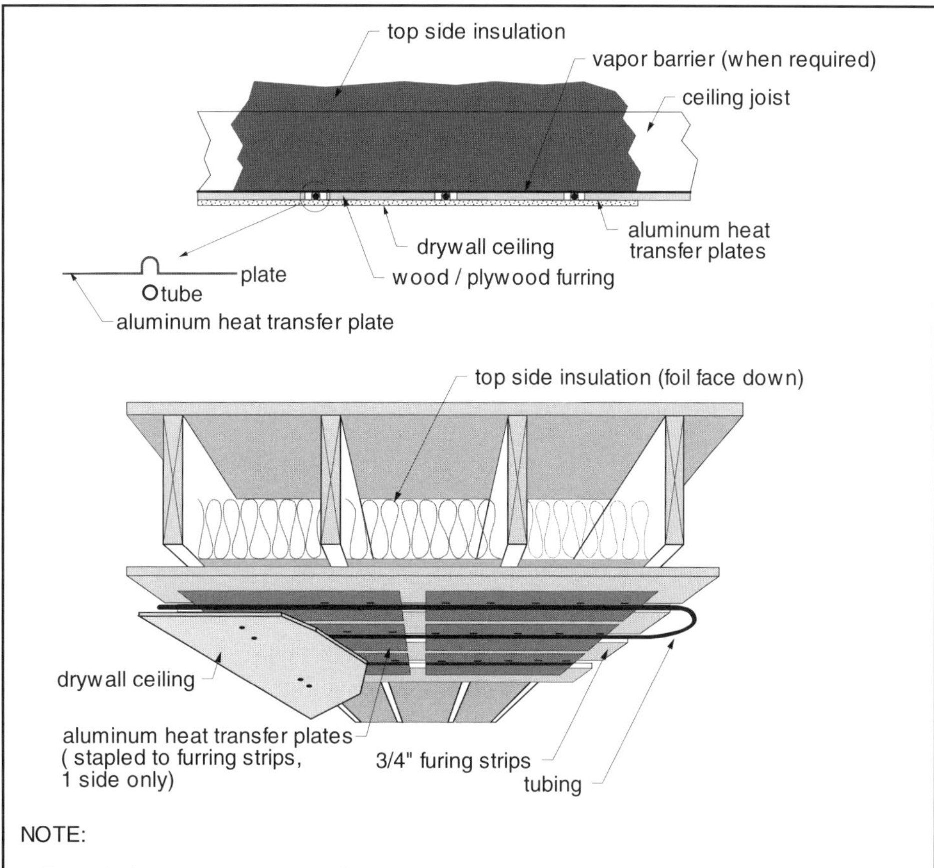

NOTE:

- Typical tube spacing: 8" to 12" (depending on heat transfer plates used)
- Maximum recommended surface temperatures:
  with 8 ft. ceiling height: 100 deg. F.
  with 9 ft. or higher ceilings: 110 deg. F.
  Gypsum panels should not be heated above 120 deg. F.
- Minimum R-value of backside insulation (with heated space above): R-11
- For ceilings beneath unheated space, increase R-value by 50% over normal ceiling R-value
- Leave minimum of 1/4" gap between adjacent heat emission plates for thermal expansion.
- Staple heat transfer plates on one side of tube only. This allows plates to lie flat as drywall is installed.
- Other top side insulation options are possible. They include reflective foil, foil-faced foam,

Figure 3-16 Radiant Ceiling Plate System

acceptable comfort to leg space under tables, desks, etc., especially if such space is directly adjacent to cold surfaces. The solution is to avoid creating shadowed floor surfaces adjacent to cold exterior walls or windows.

- **Heated ceilings usually take up less vertical space in the room.** Most hydronic radiant ceiling systems only take up 5/8" to 3/4" of room height. This is an important advantage in retrofit situations where it's often important to preserve as much room height as possible.
- **Most heated ceilings add very little weight to the structure.** Standard framing is almost always adequate to support the added weight of a hydronic radiant ceiling system.
- **Heated ceilings induce very little air circulation within a room.** Because the warmest air in the room is already against the heated ceiling, it doesn't "want" to travel downward in the room. This minimal circulation limits movement of airborne dust. It's especially advantageous in health care facilities where drafts must be avoided.

3•9 Ceiling Panel Radiators

Radiant ceiling panels have been in use within commercial buildings for several decades. They typically consist of an aluminum or steel plate that drops into a standard T-bar suspended ceiling system. Copper tubing is mechanically attached to the upper side of the panels. High-density fiberglass insulation is used to limit back and side losses. Many panels have textured or fluted surfaces that provide a simple clean appearance when installed. Such panels are easily removed or relocated if building renovation takes place. Because of their low profile, surface mounting is also possible. Figure 3-17 shows a typical T-bar ceiling panel installation.

Hydronic ceiling panels are usually installed in areas with high radiant heat loss such as above large window walls, in entry foyers, or above loading docks. They are operated at relatively high water temperatures so as to yield a high heat output from a limited panel area. Panel surface temperatures often range from 170 °F up to 185 °F. At a surface temperature of 180 °F panel heat output will be approximately 200 Btuh/sq ft.

Experience has shown that radiant heat output from these panels will also warm the floors below to temperatures higher than the room's ambient air temperature. The floor thus becomes a secondary radiating surface, further improving room comfort.

Other types of hydronic ceiling panels consist of long aluminum extrusions that are intended to mount as valances at the junction of a room's wall and ceiling. Heat is again provided by rear mounted copper tubing that is usually operated at high water temperatures.

Ceiling panels have also been successfully used for radiant cooling applications. Chilled water is circulated through the same

3

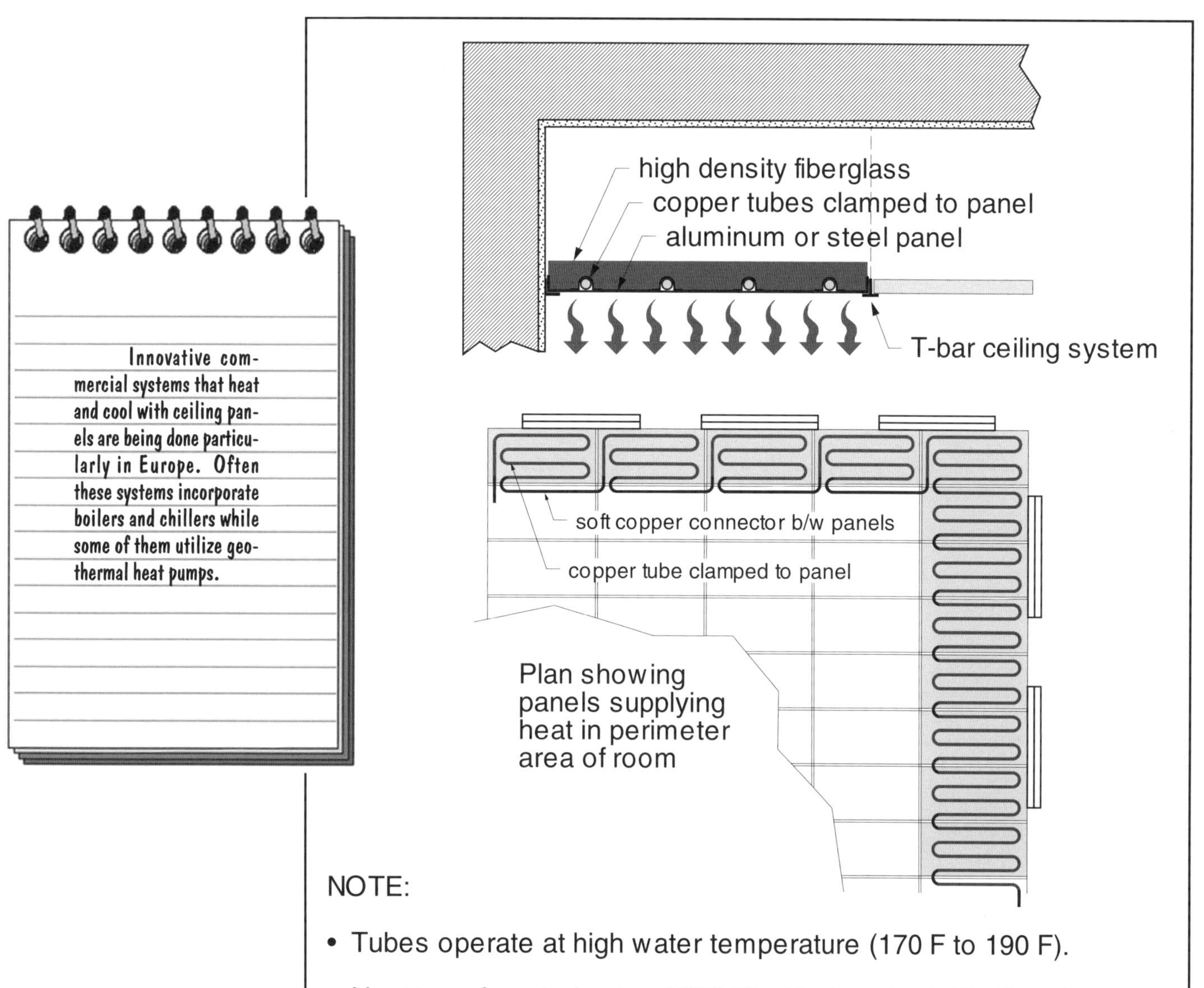

Figure 3-17 Ceiling Panels for T-Bar Ceilings

tubing to cool the panel and thus allow it to absorb heat from the room. The ceiling location is ideally suited for radiant cooling. In such applications it's crucial that the panel temperature is always maintained above the dewpoint temperature of the room. Otherwise, moisture from the surrounding air will condense on the panel.

3•10 Wall Heating

Hydronic tubing can also be installed in walls. There are two methods of installation. The first is much like a plate-type heated ceiling system. Spaced furring strips are attached to the wall framing. Aluminum heat transfer plates and tubing are installed in the gaps between the furring strips. Tubing circuits are usually routed horizontally with flow from the bottom of the wall toward the top to allow air to be effectively purged. The wall can be finished with standard drywall, skim-coat plaster, or partially faced with ceramic tile. The latter approach usually requires the installation of a cement backer board over the plates to serve as a stable substrate for the tile. Such an installation is particularly appropriate in bathrooms, swimming pool enclosures, and other areas where the wall may get wet from time to time. Figure 3-18 depicts a typical plate-type radiant wall installation.

Tubing can also be installed in a wall using a system of wood lathe, metal or polymer mesh, and layers of plaster. The lathe provides a structural base for the plaster. The mesh provides tensile reinforcement for the plaster. This approach embeds the tubing into a thermal mass that conducts heat away from the tube much like in a thin-slab floor system. No heat transfer plates are used. Such "wet" wall systems are quite common in Europe.

Drilling or nailing into walls that contain tubing obviously has to be done with care. The tubing can usually be located quite easily within a few minutes after start-up using an infrared thermometer that measures surface temperature. When PEX-AL-PEX tubing or tubing with a tracing wire is used, it can be located electronically.

When an exterior wall of a room will be heated, the R-value of its insulation should be increased by a minimum of 50% to reduce outward heat loss. Remember that the wall insulation "feels" the higher operating temperature of the tubing and plates rather than the room temperature, hence the need for higher R-value insulation. A vapor barrier will often be required behind the furring strips in exterior walls. In such cases be sure the vapor barrier can withstand the operating temperature of the tubing and plates.

Output from a heated wall is partially radiant and partially convective. The convective component is enhanced by airflow up the vertical surface. Such upward airflow can be very effective in counteracting drafts from large window areas above the heated wall.

Wall heating is also relatively easy to retrofit to an existing wall. It typically adds from 1.25" to 1.5" to the wall's thickness depending on the finish material. Electrical junction boxes will have to be fitted with extension rings to accommodate the new wall thickness.

3•11 Wall Panels

A large variety of manufactured panel radiators are available for hydronic heating applications. Unlike older cast-iron radiators, these modern panels typically only extend 1" to 3" out from the wall surface. They're available in dozens of different sizes. So-called "horizontal panels" are installed with their long dimension horizontal. They are ideal for installation under windows. Their output is a combination of radiant and convective heat. Panels without rear mounted fins tend to have

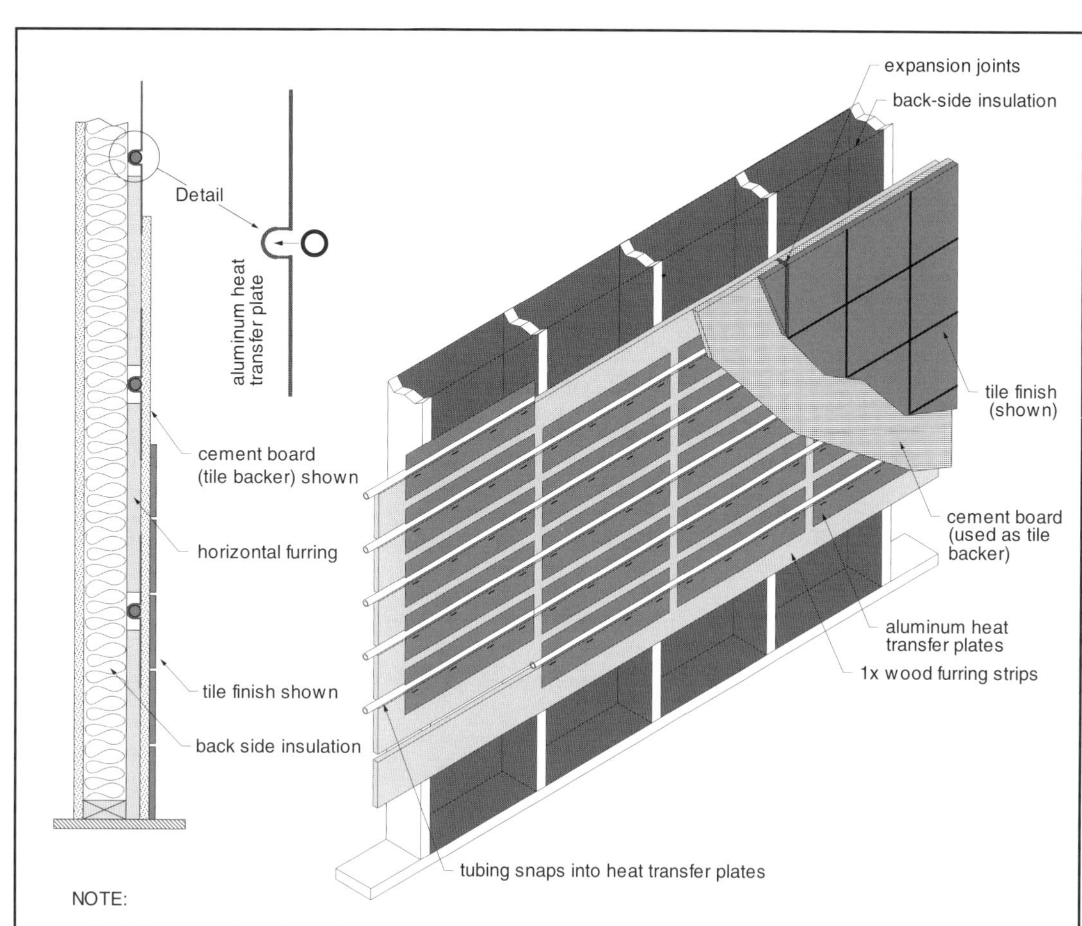

NOTE:

- Typical tube spacing: 6" to 12"
- Do not heat gypsum wall board above 120 deg. F.
- Minimum R-value of backside insulation (for interior partitions): R-11
- On exterior walls increase the R-value of back side insulation 50% above normal R-value of exterior walls. A vapor barrier is usually required on exterior walls.
- Other back side insulation options are available. These include reflective foil, foil-faced foam, and loose fill insulation.
- Leave minimum of 1/4" gap between adjacent heat emission plates for thermal expansion.
- Staple heat transfer plates on one side of tube only. This allows plates to lie flat as wall board is installed.
- Flow moves from bottom of wall up for effective air purging
- Install tile and backer board in accordance with T.C.I. recommendations.

Figure 3-18 Radiant Plate Wall System

3

higher percentages of radiant heat output.

Vertical panels can be ordered to fit wall spaces that would be difficult if not impossible to use with other heat emitters. For example, a narrow wall space in a bathroom or kitchen, perhaps only a foot wide, could be fitted with a narrow but tall panel radiator. Piping connections are typically at the bottom of the panels, although some manufacturers offer several connection options.

Most panel radiators are constructed of steel and intended for use in closed hydronic systems. Panel radiators are well-known for their high quality/durable finishes.

Most panel radiator manufacturers offer several standard (stock) panel sizes and colors. They also build custom panels for a variety of applications such as curved or segmented walls. Some panel radiators are available with built-in thermostatic radiator valves that allow for individual room temperature control.

Panel radiators can operate over a wide range of water temperature. Their heat output is strongly dependent on water temperature. When used in combination with a lower temperature heat source, such as a hydronic heat pump, a panel will have to be considerably larger to yield the same heat output that a smaller panel could deliver if operated from a conventional boiler at, say, 180 °F

Panel radiators can serve as the sole heat emitter in a room or as a means of providing supplemental heat output to other types of radiant panels under severe heating load conditions. They can be connected into series or parallel (2-pipe) hydronic distribution systems. Another method of installation is referred to as a "home run" system. Each panel is supplied from a central manifold system much like radiant floor panel circuits. This allows each panel to be separately valved and controlled on a room-by-room basis. This concept is shown in Case Study #3 in Section 2.

Another variant of panel radiators is "radiant baseboard." It consists of an extruded aluminum panel approximately 5" high and 1" thick. The panels are available in several lengths from 2 feet to 10 feet and are intended to replace typical wood baseboard. In many cases they are installed on each wall in the room. The rear side of the aluminum extrusion supports two tubes made of either copper or PEX tubing. Hot water routed through the tubes quickly heats the aluminum baseboard which in turn radiates heat into the room. Radiant baseboards can also be zoned on a room-by-room basis and supplied from a manifold system as described above. They are known for their ruggedness, particularly in commercial environments.

Components and Installation Methods

4•1 Introduction

This section examines all of the major hydronic components that go into a modern hydronic radiant panel heating system. These include the tubing, manifolds, heat source, circulator, and controls. Understanding the function and placement of each of these is crucial to achieving a stable, trouble-free, and energy-efficient system.

Although each of these components is discussed individually, it is important to remember that they ultimately function together as a system. Good design involves more than just selecting a favorite model and brand for each component. It requires that each component be properly sized so that the resulting system can maintain stable operation under varying load conditions and deliver just the right amount of heat when and where it is needed.

4•2 Tubing Options

There are several types of tubing used for hydronic radiant panel systems in the USA. These include cross-linked polyethylene (PEX), aluminum/PEX composite tubing (PEX-AL-PEX), synthetic rubber tubing, and polybutylene tubing. Copper, steel, or wrought-iron piping are rarely, if ever, used for embedded applications, although they are used for distribution piping.

Oxygen Diffusion:

Tubing made solely of polyethylene, polybutylene, or rubber compounds differs from metal pipe in one obvious respect. The molecular structure of thermoplastics and rubber compounds can allow oxygen molecules to slowly pass through the tube wall and enter the water in the system. This process is called oxygen diffusion. If not properly counteracted, it can lead to very serious corrosion problems in hydronic systems.

The reason oxygen molecules want to move through the tube wall is because the concentration of dissolved oxygen in the system's water is often low in comparison to the environment outside the tubing (even for tubing embedded in concrete). The dissolved oxygen present in the water when the system is first filled quickly reacts with iron or steel components. A slight corro-

sion film forms on these components, using up this dissolved oxygen. The water is now in a so-called "dead" state, and will not cause further corrosion in the system. However, the "oxygen vacuum" now present in the system continually tries to pull more oxygen molecules into the system through any possible path. If allowed in, the oxygen will perpetuate the corrosion of any iron and steel components in the system. The end result will be sludge buildup and the likelihood of premature failure of major system components.

The rate of oxygen diffusion varies for different materials. It also varies with the temperature of the tubing. Higher temperatures generally allow faster rates of oxygen diffusion. Tubing manufacturers can supply data on the rate of oxygen diffusion for their products. The values given generally refer to how many milligrams of oxygen will diffuse into a liter of system water (contained with the tubing) per day. To put those numbers in perspective, an oxygen diffusion rate of 10 milligrams per liter of system water, per day, is equivalent to draining and refilling the system with fresh water every single day. Experience has shown that hydronic systems containing steel or cast iron components that experience such high water turnover rates will experience swift and serious corrosion problems.

Because the tubing circuits in a hydronic radiant panel system often represent the majority of the system's volume, they provide a large potential pathway for oxygen entry. The solution to this potential problem is to create an oxygen diffusion barrier in (or on) the tubing. One such barrier is a thin layer of a special compound called EVOH (ethylene vinyl alcohol) that is bonded to the tubing during its manufacturing. The EVOH layer greatly reduces the ability of oxygen to diffuse through the tubing. Another type of oxygen barrier is a thin layer of aluminum sandwiched between layers of PEX (PEX-AL-PEX) tubing. An internationally recognized standard for the performance of oxygen diffusion barriers is known as DIN 4726. Tubing that meets this standard allows no more than 0.1 milligrams of oxygen to enter the system per liter of water in the system, per day (based on 104 °F water temperature). This amount is considered low enough to prevent any detrimental corrosion. It is highly advised that any non-metallic tubing being considered for use in hydronic systems which contain any cast-iron or steel component meet the DIN 4726 standard.

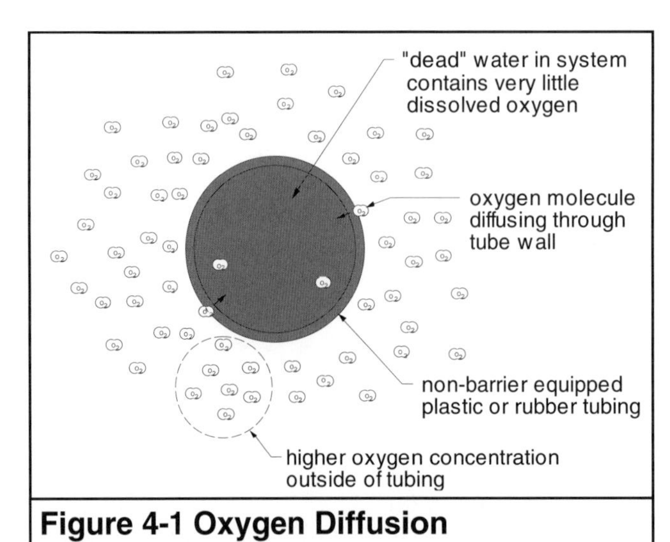

Figure 4-1 Oxygen Diffusion

If non-barrier tubing is used in a hydronic radiant panel system, ALL system components must be made of non-corrodible materials such as stainless steel, copper, bronze, brass, or thermo-

plastics. Water quality must also be tested to ensure that the presence of free oxygen in the water will not support any chemical reactions with these materials, nor sustain bacteria or other microorganisms. Any such systems should also be clearly labeled as containing non-barrier tubing and any restrictions this implies should be clearly stated to ensure that iron or steel components will not be used as replacements in the future.

One final point concerning oxygen entry into hydronic systems needs to be stressed. There are several ways other than diffusion through non-metallic tubing by which oxygen can enter a hydronic system. These include improperly sized or placed expansion tanks, leaky valve seals or pump gaskets, and improperly located air vents. All such deficiencies have the potential to allow small but frequent quantities of fresh (oxygen-containing) water or air into the system. The resulting corrosion could be as bad or worse than that resulting from diffusion through non-metallic tubing. In other words, the use of an oxygen barrier-equipped tubing does NOT guarantee that oxygen-related corrosion will not occur. However, the slight additional cost of providing barrier-protected tubing is small in comparison to the repair costs associated with extensive corrosion.

PEX Tubing:

Cross-linked polyethylene tubing (PEX) is currently the most widely-used tubing in hydronic radiant panel systems. Each year hundreds of millions of feet of PEX is installed worldwide. However, many American tradesmen are not yet familiar with this product. This stems from the fact that PEX tubing was developed in Europe, making its first commercial appearance in the USA in the early 1980's. Some of the PEX tubing currently sold in the US is still produced in Europe.

Polyethylene, the base material of PEX tubing, is derived from crude oil. It consists of large chains of carbon and hydrogen atoms joined through a process called polymerization. At present, polyethylene is used for thousands of different products such as food containers and automobile parts. Cross-linking is what distinguishes PEX from ordinary polyethylene. In this process, neighboring polyethylene molecules are bonded together to form a three-dimensional chemical structure. In a matter of speaking, a coil of PEX tubing can be considered one giant molecule. Cross-linking changes the polyethylene from a simple thermoplastic material into a thermoset plastic. After cross-linking, the shape of the polyethylene is fixed. It cannot be changed back to a liquid and re-formed, even at high temperatures.

Cross-linking improves both the temperature and pressure ratings of PEX tubing relative to ordinary high-density polyethylene tubing. PEX tubing that carries the ASTM F876 rating is capable of transporting 180 °F water at 100 psi. It also is rated for service at 200 °F at 80 psi. These ratings apply to continuous service conditions and contain built-in safety factors. During manufacturing, PEX tubing is tested to temperatures and pressures significantly higher than these ratings. One manufacturer reports continuous testing of their PEX tubing at 200 °F and 170 psi, for over twenty years, with no detrimental loss of strength.

Nearly all hydronic radiant panel heating systems operate at temperatures and

pressures well below these rating conditions. For example, a typical slab-on-grade floor heating system will likely operate with water temperatures of 100 to 110 °F with corresponding pressures in the range of 15 to 20 psi. Because such systems put relatively low service demands on the tubing, its expected life increases significantly. Exactly how long PEX tubing will last is hard to say. But given the service conditions under which the tubing operates in many radiant panel installations, it will most likely outlast the building in which it's installed.

PEX tubing also has a "shape memory" characteristic that allows it to be restored to its original shape after being kinked. This process is activated by heating the PEX to about 275 °F. At this temperature, the normal crystalline structure of PEX changes to an amorphous state and the tube's original shape is restored. This allows an accidental kink in PEX tubing to be repaired on the job site.

At room temperature conditions, heating the kinked area to the amorphous state will take about two minutes using a standard industrial hot air gun. The kinked area should be heated uniformly. Never attempt to heat the tubing using a flame. As the kinked area reaches the amorphous condition, the characteristic milky-white color of PEX changes to transparent. The tubing will now have returned to its original shape. Allow the repaired area to cool at its own rate for at least 5 minutes before handling the tubing. A slight blistering effect is often noticeable on the surface of oxygen barrier PEX after the repair. This is caused by distortion of the EVOH oxygen diffusion barrier layer and is normal.

Most PEX tubing produced in Europe is manufactured in metric sizes. Under the European system, tube size refers to the outside diameter (O.D.) of the tubing in millimeters, rather than the nominal inside diameter used in the USA plumbing trade. Typical USA tube sizes used in residential and light commercial radiant panel systems are shown in Figure 4-2, along with the approximate equivalent metric dimensions.

The smaller tube sizes are used mainly in residential radiant panel systems, while the larger sizes are used in commercial and industrial systems.

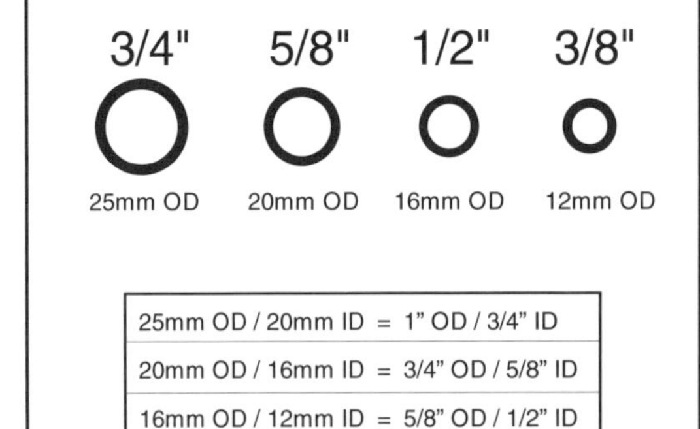

25mm OD / 20mm ID	= 1" OD / 3/4" ID
20mm OD / 16mm ID	= 3/4" OD / 5/8" ID
16mm OD / 12mm ID	= 5/8" OD / 1/2" ID
12mm OD / 8mm ID	= 1/2" OD / 3/8" ID

- 3/4" and 5/8" tubing are more common in commercial systems. 3/4" is common in sno' melting applications.
- 1/2" and 5/8" tubing are more common in residential systems.

Figure 4-2 Typical Tubing Sizes

PEX-AL-PEX Tubing:

In recent years another type of tubing has become increasing popular for use in radiant panel heating systems. It consists of a "core" of ultrasonically welded aluminum with inner and outer layers of cross-linked polyethylene. The PEX material is bonded to the aluminum core with special adhesives as the pipe is extruded.

PEX-AL-PEX tubing used in radiant panel applications should meet the ASTM F1281 standard. Temperature ratings of 200 to 210 °F at pressures of 100 to 115 psi are typical (check ratings from individual manufacturers).

The aluminum core of PEX-AL-PEX tubing provides an excellent oxygen diffusion barrier, easily meeting the DIN 4627 standard. This core also allows the tubing to retain the shape once formed. Accidental kinks in PEX-AL-PEX tubing are repaired using a reshaping tool to reform the kinked area to its original shape.

PEX-AL-PEX tubing used for radiant panel heating systems is typically 1/2", 5/8", or 3/4" nominal size.

A variety of fittings are available to adapt PEX-AL-PEX tubing to standard N.P.T. threaded hardware. Specialized "press fittings" are also available.

Polybutylene Tubing:

Another thermoplastic material used for hydronic tubing is polybutylene (or PB for short). PB tubing is manufactured from a plastic resin derived from crude oil. It is available in a number of sizes and coil lengths. Only PB tubing that carries the ASTM D-3309 specification with its corresponding 180 °F at 100 psi rating should be used in radiant panel applications. Polybutylene tubing is currently available with or without an oxygen diffusion barrier. Barrier equipped tubing should meet the DIN 4726 standard for oxygen diffusion. Unlike PEX, PB tubing is not cross-linked. This does not allow it to obtain the higher temperature/pressure ratings of PEX. However, this is seldom a problem in most radiant panel applications and floor heating applications that operate at temperatures and pressures significantly below these ratings.

Because it is not cross-linked, PB tubing does not have a shape memory, and cannot be repaired using the heating process previously described for PEX tubing. If a kink occurs during installation, the affected area must be cut out of the tubing and repaired. PB tubing without an oxygen diffusion barrier can be repaired using either a mechanical coupling or through a process called socket fusion. The latter involves heating the outside of the tube wall and the inside of a fitting to just over 500 °F using a special tool. This softens the PB into a semi-molten, but relatively undeformed, state. The tube and fitting are then quickly pushed together. As they cool, the tube and fitting fuse together. The resulting joint is very strong and permanent. Polybutylene tubing that has an oxygen diffusion barrier cannot be joined using socket fusion. In such cases, a mechanical coupling is the only alternative. When such a repair is made it's advisable to leave access to the joint in case the coupling has to be tightened in the future. In general, care should be taken during installation to ensure the tubing is not severely kinked. Gentle and slow bending, combined with a smooth uncoiling system, will guard against kinking.

Rubber Tubing:

Radiant panel systems can also be installed using tubing made from synthetic rubber compounds. Since rubber is inherently more elastic than either polyethylene or polybutylene, a layer of reinforcing mesh is incorporated into the tubing to enable it to obtain suitably high pressure/temperature ratings.

Rubber tubing has the advantage of being very flexible in comparison to other polymer tubing. This enables it to be easily routed around potential obstacles such as plumbing pipes or structural columns. It also makes the tubing extremely resistant to kinking or crushing during installation.

Like PEX and PB, rubber-based tubing is available with or without an oxygen diffusion barrier. The molecular structure of rubber compounds allows oxygen molecules to pass through tube walls at a greater rate than either polyethylene or polybutylene. For this reason, it is essential to use the barrier-equipped tubing if there are any steel or iron components in the system. As with PEX and PB, barrier-equipped tubing should meet the DIN 4726 standard if used for radiant panel heating applications.

Comparison of Tubing Options:

1. Cross-linked Polyethylene (PEX) Tubing

ASTM F876-93 Standard Methods for Cross-linked Polyethylene (PEX) Tubing

ASTM F877-93 Standard Methods for Cross-linked Polyethylene (PEX) Plastic Hot- and Cold-Water Distribution Systems

(PEX tubing is available with and without an oxygen diffusion barrier)

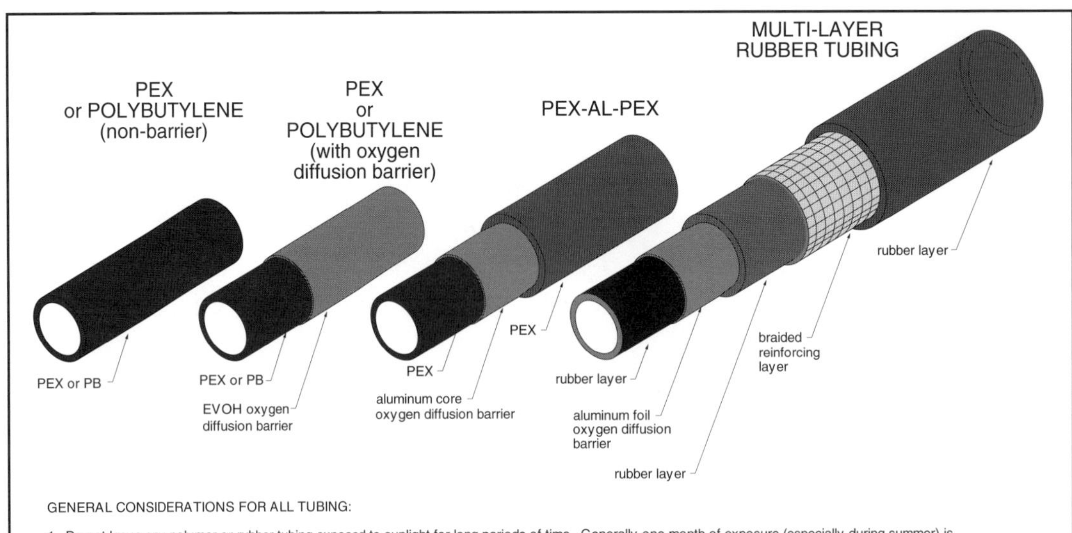

GENERAL CONSIDERATIONS FOR ALL TUBING:

1. Do not leave any polymer or rubber tubing exposed to sunlight for long periods of time. Generally one month of exposure (especially during summer) is considered maximum, although manufacturers differ on recommendations.

2. Always use fasteners that are recommended by the tubing manufacturer. In some cases the tube's warrenty is dependent on use of approved fastners.

3. Do not exceed the manufacturers specified operating conditions for temperature and pressure.

4. Do not bend the tubing into tigher radii than recommended by the manufacturer.

5. Always uncoil tubing from its roll . NEVER pull the tubing off the side of its roll.

6. Although all tubes are resistant to damage during construction, use reasonable protection whenever appropriate. Do not chop into the concrete with shovels during a pour. Do not drag objects over tubing that could abraid its surface. Be especially careful of heavy equipment such as pipes for pumped concrete, "powerbuggies", concrete truck shoots, and so forth.

7. Always use tubing with an oxygen diffusion barrier in systems containing iron or steel components.

8. Do not attach tape (duct tape) to any type of PEX tubing. Solvents in the adhesive may damage the PEX over time.

Figure 4-3 Tubing Options

CSA B137.5
Service ratings: 180 °F @ 100 psi and 200 °F @ 80 psi (PEX tubing is available with and without an oxygen diffusion barrier)

2. (PEX / Aluminum / PEX) Composite Tubing

ASTM F1281-96 Cross-linked Polyethylene/Aluminum/Cross-linked Polyethylene (PEX-AL-PEX) Pressure Pipe
CSA B137.10
Service ratings: 73 °F @ 200 psi, 140 °F @ 150 psi, 180 °F @ 125 psi, 210 °F @ 115 psi (refer to individual manufacturer's rating data)

3. Rubber Tubing

ASTM D380-94 Rubber Hose
ASTM D413-82 (1993) Rubber property - Adhesion to Flexible Substrate
ASTM D471-96 Rubber Property - Effect of Liquids
ASTM D395-89 Rubber Property - Compression Set
ASTM D412-97 Vulcanized Rubber & Thermoplastic Rubber & Thermoplastic Elastomers - Tension
Service rating: 210 °F @ 100 psi

4. Polybutylene (PB) Tubing

ASTM D3309-96a Polybutylene (PB) Plastic Hot- and Cold-Water Distribution Systems
CSA B137.8
Service ratings: 180 °F @ 100 psi
(PB tubing is available with and without an oxygen diffusion barrier)

Oxygen barrier rating:

DIN 4726 allows a maximum of 0.1 milligrams of oxygen entry per liter of water in tubing per day at a water temperature of 40 deg. C. (104 °F)

4•3 Circuit Layout Considerations

The placement of tubing circuits determines how well the hydronic radiant panel system operates and what degree of comfort it provides. There are several things that have to be considered simultaneously, and of necessity, most layouts are usually compromises between several "competing" factors. These include:

- Allowing for room-by-room temperature control.
- Placing the warmest portion of the circuit adjacent to cooler exterior walls.
- Ensuring there is sufficient tubing within the space to properly heat it.
- Ensuring that variations in panel surface temperature don't cause complaints.
- Placing tubing to accommodate changes in flooring materials.
- Determining if the building requires different water temperatures in different circuits.
- Avoiding excessive circuit length.
- Minimizing locations where tubing crosses control and/or construction joints.
- Minimizing the number of return bends (or other tight bends) in the circuit.

- Minimizing “leader” distance between manifold and area served by circuit.
- Avoiding areas where the tubing is subject to damage from drilling, sawing, or fasteners.
- Avoiding tubing overlaps.
- Providing higher heat output in areas subject to tracked-in moisture (such as entry areas).
- Determining areas/rooms in the building that will require more than one circuit.
- Having to run tubing parallel to, and/or within the joist cavity.
- Maintaining equal length circuits in certain situations.

Some of these considerations can only be addressed after making the calculations presented later in the manual. Others tend to be more consistent from one building to the next and can be handled by the following rules of thumb.

Circuit Layout Patterns:

Most circuit layout patterns fall into one of the following groups:

- One-directional serpentine pattern
- Two-directional serpentine pattern
- Three-directional serpentine pattern
- Counterflow spiral pattern

The three serpentine patterns allow the warmest part of the circuit to be routed near the cooler surfaces in the room (the exterior walls, windows, and doors). This creates slightly higher rates of heat output in areas that tend to increase the body’s rate of heat loss. The result is improved comfort. Furthermore, in the case of floor heating, the warmer floor areas create a gentle upward airflow that counteracts downward drafts from the cooler exposed surfaces.

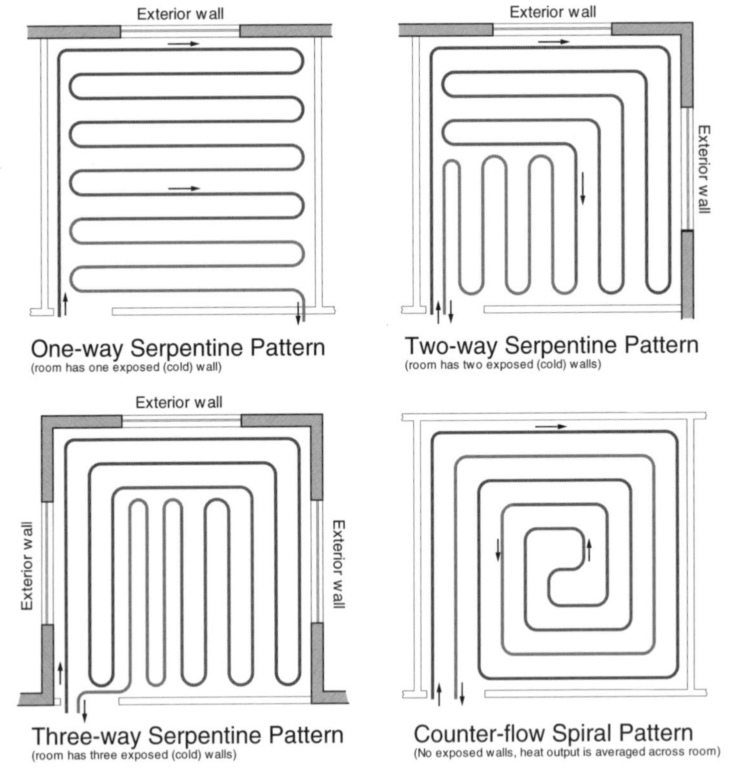

NOTES:

1. Note flow direction arrows on circuits. Warmest water entering circuit is routed adjacent to exterior walls so as to yield higher heat output in cooler portin of room.

2. Maintain 4” to 6” spacing between all walls and tubing to minimize chances of accidental damage due to drilling, sawing, or nailing.

3. To prevent puncture damage, accurately mark any locations where tubing circuits are routed beneath partitions. Avoid driving any fasteners into floor within 6” on either side of tubing.

Figure 4-4 Circuit Layout Patterns

The counterflow spiral pattern differs considerably from the serpentine patterns. This pattern does not "bias" heat output toward any given area of the room. Rather, it attempts to average heat output over the entire area. It's a good choice for totally interior rooms, or large interior areas where multiple circuits are required but heat output needs to be equally distributed. It does have the advantage of requiring mostly 90 degree bends rather than 180 return bends. This allows closer tube spacing where necessary. This pattern also "interlaces" the warmer portions of the circuit with the cooler portions, resulting in less variations of floor surface temperature in comparison to a serpentine pattern.

Variable Tube Spacing:

Some radiant system designers prefer to use closer tube spacings in areas of rooms that may have exceptionally high heat losses. These would include floor or ceiling areas adjacent to large "window walls" or walls with relatively low R-value. The closer spacing allows greater heat output to compensate for lower mean radiant temperatures. In floor areas where prolonged foot contact will not occur, the surface temperatures can be allowed to climb into the lower 90 °F range. This will let heat output approach 50 Btuh/sq. ft. These areas of high heat output usually only extend 2 to 4 feet in from the exterior walls. Closer tube spacing is also sometimes used in "transitory" areas such as entry foyers and halls to speed drying of floors subject to tracked-in snow and water. The tubing layout plan in Figure 4-5 illustrates closer tube space adjacent to exterior log walls that have relatively low R-value and hence lower inside surface temperatures.

Tubing Circuit Lengths:

The length of tubing circuits, whether installed in floors, walls, or ceilings, will directly affect the flow rate a given circulator can produce through the circuits. This in turn affects heat output as well as surface temperature variation of the radiant panel. If a circuit is too long, its low flow rate will limit heat output as well as exaggerate variations in surface temperature, both of which are undesirable. Once installed, such "mistakes" are very difficult and expensive to correct.

The following maximum circuit lengths are recommended by the RPA Standard Guidelines:

- 1/4" (nominal ID) tubing: 100-125 ft.
- 3/8" (nominal ID) tubing: 200-250 ft.
- 1/2" (nominal ID) tubing: 250-350 ft.
- 5/8" (nominal ID) tubing: 400-500 ft.
- 3/4" (nominal ID) tubing: 500-600 ft.
- 1" (nominal ID) tubing: 750 ft.

These lengths were established in consideration of the typical small wet-rotor circulators used in most systems, as well as what are acceptable variations in the surface temperature of the radiant panel. The range of length given for each tube size allows for both "conservative design," as well as "acceptable maximums" in cases where variation in the surface temperature is not as critical (a garage floor for example). These lengths include not only the tubing within the room, but any "leader" lengths between the room and manifold location as well.

Equal Length Circuits:

Some hydronic radiant panel systems have been designed and installed with great care given to achieving circuits of equal length. The goal is to ensure equal flow rates in each circuit. Although it's true that when circuits of equal length and tube size are all connected to the same manifold, their flow rates will be equal, it is not always necessary to achieve this. The "goal" of the system is to provide the right amount of heat where it's needed. When circuits are laid out on a room-by-room basis and flow adjustment valves are provided on the return manifold(s), this goal can be achieved without having equal circuit lengths and equal flow rates. The balancing valves provide a way to induce additional heat loss in short circuits and thus prevent disproportionately high flow rates. This allows heat output to be balanced as needed.

There are, however, cases where circuits of approximately equal length are desirable. For example, when a large floor area requires several circuits and that entire area will be controlled as a single zone, then circuits of approximately the same length can eliminate the need for individual flow balancing valves on the return manifold(s). This will reduce cost, especially in large systems. Keep in mind, however, that in multiple manifold systems, keeping the circuits all the same length still does not guarantee all circuit flows will be the same. The distribution piping serving the multiple manifold system will create different pressure differentials across each of the manifold stations, which in turn affects circuit flow rates. In such cases balancing valves on each manifold station (but not necessarily each circuit) are advisable.

Rooms with Multiple Tube Circuits:

Given the limit on tubing circuit length and typical tube spacings of 6 to 18 inches, it's not uncommon for some rooms to require two or more circuits to cover their floor area. In such cases the circuits should either be of approximately the same length, or each circuit should be provided with a balancing valve on the return manifold.

Some multiple circuit rooms are designed with a "perimeter circuit" that has closer tube spacing and operated at a high water temperature to concentrate heat output in areas adjacent to cool exterior walls, window, or doors. Other circuits with wider tube spacing, and/or lower water temperatures, serve interior areas of the room. Before using this concept with a floor system, however, the designer should confirm that the perimeter area will not be heavily obstructed with furniture or other object that could significantly restrict heat output.

Tubing Layout Plan:

Designers of hydronic radiant panel systems are encouraged to make a scaled tubing layout plan for every installation. Although an accurate drawing takes time to construct, the benefits are well worth it. They include:

- Providing the installer with a detailed plan of how the tubing should be installed. This avoids misunderstandings, "trial & error" installation methods, incorrect coil cuts, or other results that differ from those expected by the designer.
- Much faster installation is possible with an accurately-drawn plan. In many cases the time spent drawing the tubing layout plan is more than recovered by reduced installation

time, especially for new installers.

- Circuit lengths can be accurately determined before tubing is purchased or installed. This avoids excessive circuit length, as well as finding that the tubing segment being installed is too short to make it back to the manifold after the majority of it has been fastened in place. It also allows more detailed quantity estimating, as well as "optimization" of how individual circuits should be cut from both existing and new coils.
- The plan provides a permanent/accurate record of where the circuits are installed, and what manifold station they are connected to. This can save significant time and frustration during balancing, service, or when building renovation is performed.
- It provides other trades on the job, (both present and future) with information that can prevent damage to embedded tubing circuits from "blind" drilling, sawing, etc. In nearly every case accidental damage to embedded tubing occurs because someone didn't realize it was there.

Making the Plan:

Although it's possible to manually draw a tubing layout plan, the availability and relatively low cost of computer-aided-

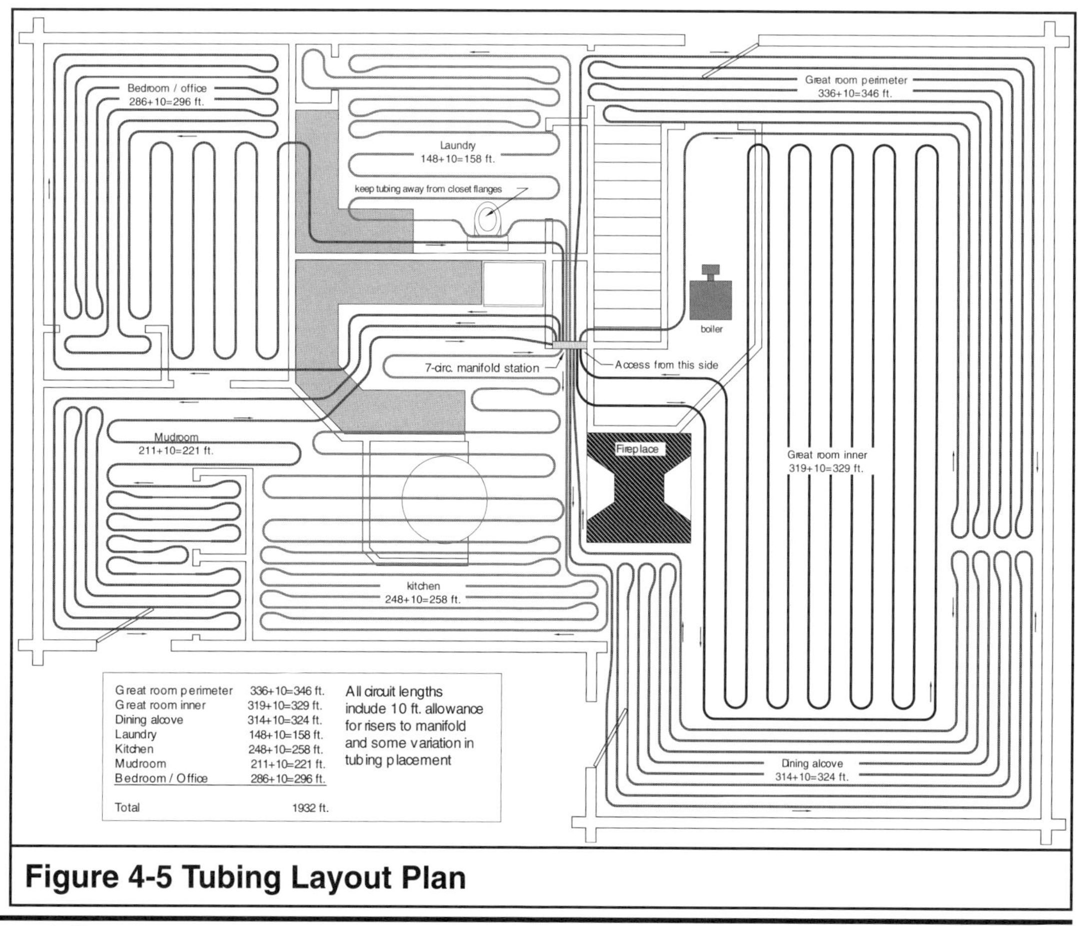

Figure 4-5 Tubing Layout Plan

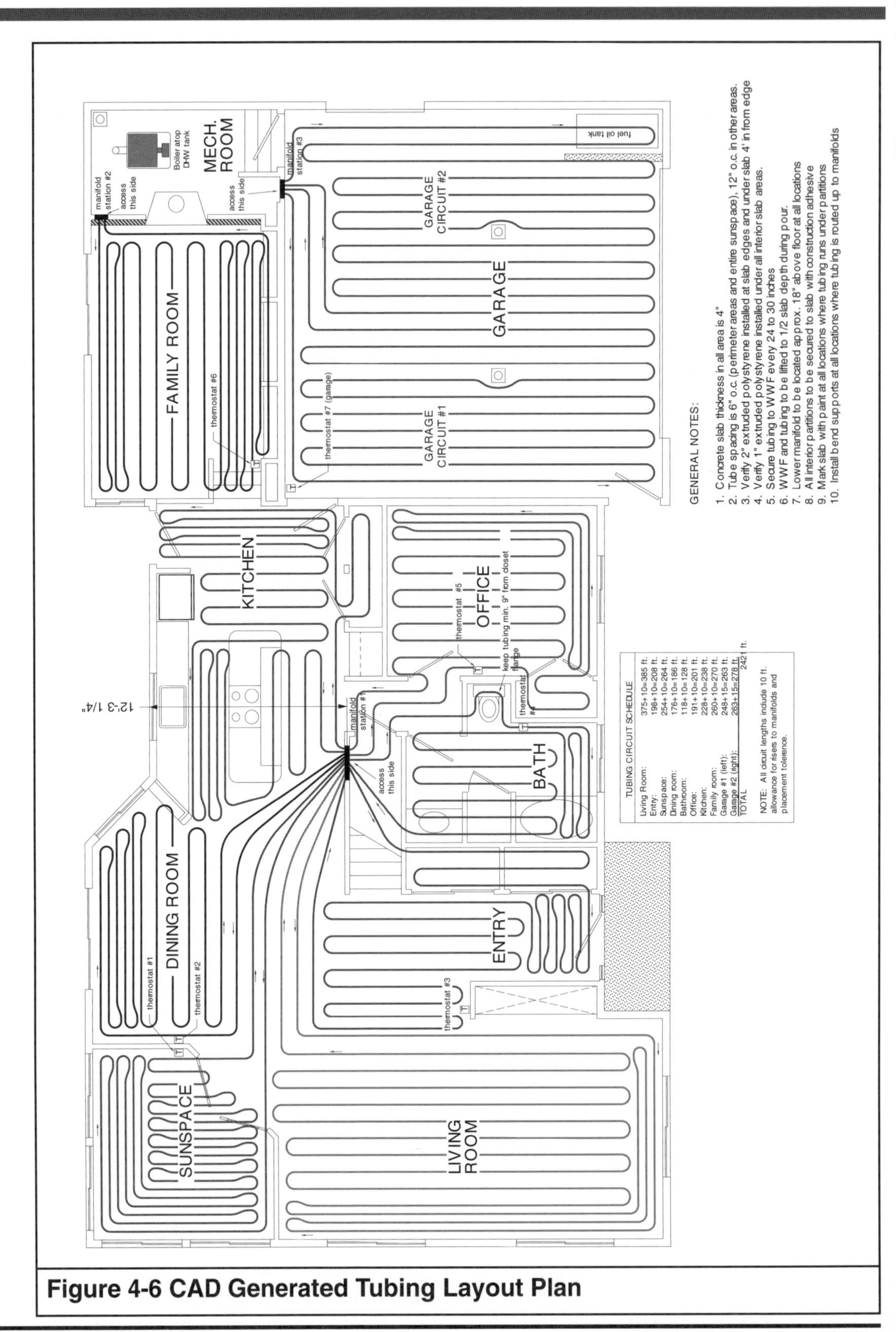

Figure 4-6 CAD Generated Tubing Layout Plan

drawing (CAD) programs make this the preferred method of producing the plan. Several CAD programs costing less than $300 are presently available for both Windows® and Macintosh® computers. All are capable of precision scale drawing, and most can also be configured to automatically determine the length of completed circuits. When the plan is completed, it can be printed to any number of low-cost inkjet or laser printers.

Optimal drawing techniques will vary from one CAD program to another. Before deciding on a particular program, the designer should contact the software vendor and describe exactly what it is that the program needs to do. The vendor should then be able to recommend a specific drawing technique for their software.

Most drawings would begin by establishing an accurate "'template" of the building's floor plan. Although this template needs to be drawn to accurate scale (to allow the program to accurately measure circuit lengths) it does not have to include all the details shown on a typical floor plan. Remember that the drawing's purpose is for placing tubing, not for locating windows, doors, receptacles, or other architectural elements.

In cases where the building plans have also been drawn using a CAD system, the building designer may be able to provide a disc file containing wall and partition locations. Many of the current CAD programs can import such a file in various formats such as DXF or DWG. If such files are available they can often save time in constructing the tubing layout plan, especially for complicated floor plans. It's advisable to "strip out" any irrelevant details, notes, dimensions, etc. from such CAD files before laying out the tubing circuits, since they only tend to clutter the plan. When the basic floor plan template has been established it is also advisable to draw each floor circuit on a separate "layer" in the CAD program. This allows easy modifications, deletions, etc. without disturbing other elements of the plan.

The following items are suggested for inclusion in the tubing layout plan:

- Name of project and the scale at which the drawing is printed
- Name of each room
- Exact location of each manifold station within the building (with placement dimensions)
- A name for each manifold station
- The direction from which each manifold station is accessed once walls are closed in
- The routing path of each circuit
- The flow direction in each circuit
- A name for each circuit for reference in a list (if other than room name)
- The location of all control/construction/expansion joints in slab
- A schedule listing each circuit by name and length
- The location of any thermostats used
- The location of any valves or controls other than those at the manifold station(s)
- All locations where floor is to be marked to prevent accidental damage to tubing
- Notes and/or details relevant to the installation (if not provided on other drawings)

Photograph the Completed Tubing Installation:

In addition to tubing layout plans, it's advisable to take several photographs or a video recording of the completed tubing installation. This will pick up any deviations from the plans as well as provide quick reference to tubing location if future work is done on the floor. Photographs of each side of each manifold location are especially helpful.

Manifold Stations:

Manifold stations consist of a supply manifold at which one or more tubing circuits begin and a return manifold where these circuits end. Manifold station are usually located in recessed cavities between wall studs in wood-framed buildings, or sometimes mounted directly to wall surfaces in mechanical rooms or industrial buildings.

There are many types of manifolds presently available in the US. They range from simple valveless headers fabricated from copper tubing to machined brass assemblies with accessories such as flow balancing valves, air vents, flow meters, and fill/drain valves. Some manifold systems must be ordered with a specific number of supply/return circuits. Others are modular in nature, allowing the required number of circuit connection to be assembled as needed. A given manifold station can serve from 1 to upwards of 12 individual circuits, depending on its design and the flow rates in the circuits.

All tubing suppliers have at least one type of manifold available for use with their tubing. By purchasing the tubing and manifolds from the same manufacturer, you're ensured of compatibility between the tubing and manifold connection. If you want to "mix and match" manifold parts and tubing from different suppliers, be absolutely sure to check tubing/connector size and thread type first. Some manifolds are made with standard NPT pipe threads, for standard US tube sizes, while others are imported from Europe and use metric tube sizes and euro-conical threads. Figure 4-7 shows a typical manifold station installed within a stud cavity.

Manifold Placement:

The number and placement of manifold stations in a building depends on several factors, including:

- Do all floor circuits operate at the same supply water temperature? (A given manifold station can only supply one temperature to all its circuits at one time.)
- Can all floor circuits be routed from a single manifold without excessive "leader" lengths?
- Can the flow rate of all system circuits be handled by a single manifold?
- Is each circuit independently controlled, or are several circuits controlled as a group?
- What manifold mounting locations are available within the building?
- How many floor levels does the building have?
- Will some circuits be filled with an antifreeze solution while other operate with water?

Ideally, manifold stations are placed so that circuits can be routed away from them in several directions. Think of the manifold station as the hub of a wheel, with the circuits as the spokes. Such "radial"

arrangements with several circuits clustered around the manifold station tend to minimize the length of circuit leaders (the tubing between the manifold and the rooms where the circuit is intended to release its heat).

In buildings with wide, spread-out floor plans it is usually better to install two or more manifold stations (each with circuits clustered around it) rather than attempting to route all circuits back to a single manifold location. The latter approach also tends to concentrate considerable amounts of tubing in hallways that have very little heat loss and thus are often overheated. Various distribution piping options are available for multiple manifold systems depending upon if they're controlled separately or as a group. These are discussed later in the manual.

Some radiant panel systems, especially floor heating systems installed in buildings with different types of finish flooring, require two or more manifold stations to service circuits that need to operate at different temperatures. Before designing the system it is crucial to know what supply temperature is required for each circuit, as well as where these circuits are located in the building. The idea is to "collect" circuits that operate at approximately the same supply temperature onto the same manifold station, especially when they're in the same vicinity in the building.

Once the number and general location of the manifold stations have been determined, look to the floor plan for exact mounting locations. The wall of a closet, with the access panel opening into the closet, is a good choice since few owners object to the access panel being visible from within a closet. Locating an access panel in the middle of a living room, on the other hand, may be objectionable from an aesthetic standpoint. Make sure that the location chosen allows easy access (remember it takes a while to move a 500 pound bookshelf out of the way, to make a valve adjustment). Again it pays to coordinate the exact location of the manifold access panel with the owners, and/or building designer.

Valved vs. Valveless Manifolds:

Any given manifold station can be configured with or without valves for its circuits. The best choice for a given project depends largely on how the circuits are controlled. When all circuits served by a manifold heat portions of a single large floor area and all have approximately the same length, there's usually no need for flow balancing valves on each circuit. A simple valveless (header) manifold can be used to reduce installation cost. However, when the circuits serve different rooms or have significantly different lengths, a manifold with individual flow balancing valves, one for each circuit, is advised. Such valves are typically located on the return manifold.

The valves provided on valved manifold systems are typically a combination balancing and on/off valve. They consist of a spring-loaded stem assembly that adjusts the position of the valve's plug above an orifice. Water returning from a given circuit flows directly against the plug. To reduce flow through the circuit, the stem height is adjusted by manually turning a portion of the valve's mechanism to lower the plug closer to the orifice. The spring assembly, however, always holds the

valve's plug as far open as is possible, given its balancing setting.

Some manufacturers provide computer-generated numbers for setting manifold valves to achieve the desired flow rate in each circuit. In such cases, the valves are adjusted as part of the start-up routine after the system has been filled and purged. When valve settings are not determined during system design, they are usually set by trial and error. As a rule, shorter circuits require more valve closure compared to longer circuits. Short tubing circuits tend to "steal" more flow than they need from the manifold. This usually leads to underheating in areas served by the longer circuits, and overheating in areas served by the short circuits. By adjusting the balancing valves, manifold flow can be redirected as needed by the individual circuits. When a new system is started, it's good to allow the system to operate for a few days with all balancing valves fully open (or at their initial "guesstimated" settings) and observe which areas of the building are overheated or underheated. The balancing valves can then be fine-tuned as required. Be sure to keep a written log of the setting of each balancing valve on each manifold as reference when and if future adjustments need to be made.

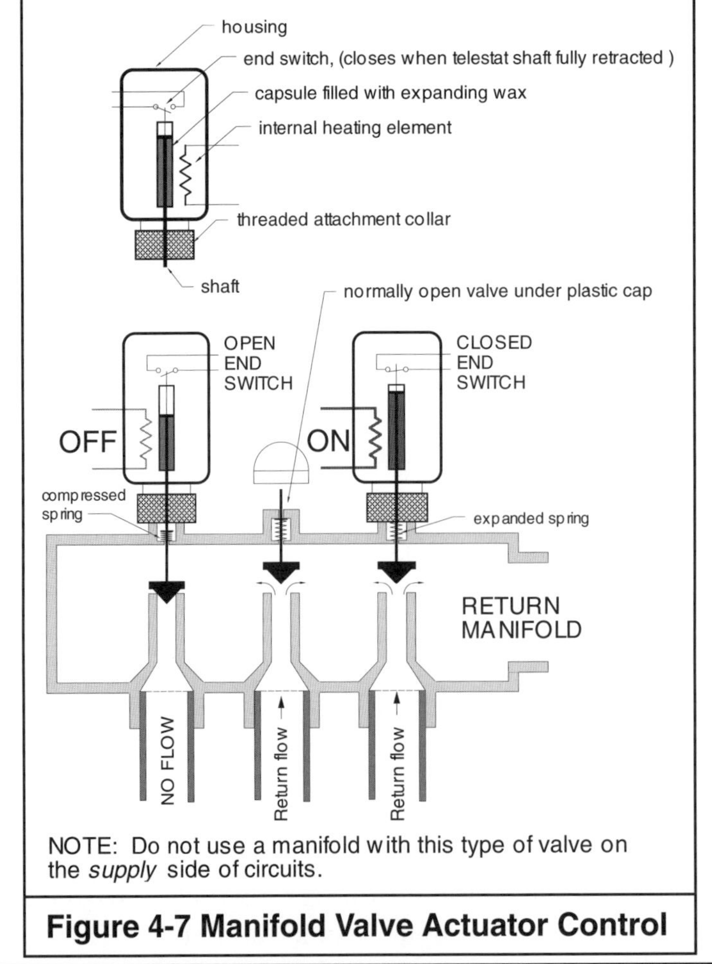

Figure 4-7 Manifold Valve Actuator Control

Valve Actuators:

Once a balancing valve has been set, the remaining stem travel is intended to be controlled by an electric valve actuator. The actuator, which is typically operated by a 24 VAC signal like a standard hydronic zone valve, screws down onto a manifold valve. As it is screwed in place, the actuator's stem pushes the spring-loaded valve stem all the way down, completely closing the valve. When a 24 VAC signal is applied to the actuator, it retracts its stem, allowing the spring-load manifold valve to reopen as far as the balancing setting allows. Flow through a given circuit can thus be turned on and off by a low-voltage control signal from a room thermostat. This is an essential part of room temperature control in many systems.

Some valve actuators use a heat motor to create the motion necessary to control the valve stem. In such devices the 24VAC signal powers a resistor that heats a fluid inside a flexible diaphragm assembly. The heated fluid expands, causing the diaphragm to change length and move the valve stem. Heat motor type valve actuators can take from 3 to 5 minutes to fully open. Generally this opening time requirement is not a problem, given the relatively slow response of many radiant panels. Other types of actuators use small gear motors to produce the stem travel. They can fully open a valve in just a few seconds. Both types of actuators have been successfully used in many radiant panel heating applications.

Valve actuators can be mounted on any one or more of the valves on a return manifold. This allows the option of individual thermostatic control of each floor circuit in applications where needed, an example being where individual circuits serve each of several rooms that the owners want to maintain at different temperatures. Another example is where solar heat gains would overheat some rooms while others rooms still require heating. In this case, the thermostat in a room experiencing solar heat gain would sense the increased temperature and respond by stopping flow in the associated tubing circuit. When the solar heat gain diminished and the room's temperature started to drop,

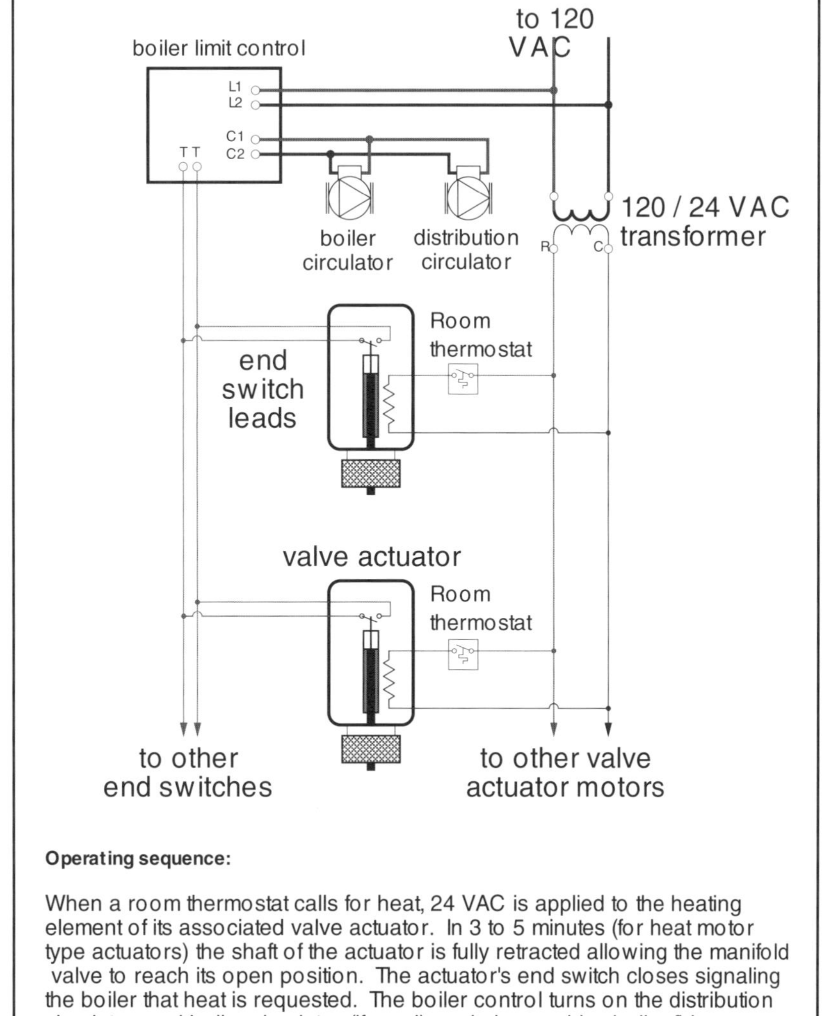

Operating sequence:

When a room thermostat calls for heat, 24 VAC is applied to the heating element of its associated valve actuator. In 3 to 5 minutes (for heat motor type actuators) the shaft of the actuator is fully retracted allowing the manifold valve to reach its open position. The actuator's end switch closes signaling the boiler that heat is requested. The boiler control turns on the distribution circulator, and boiler circulator (if used), and also enables boiler firing.

Figure 4-8 Valve Actuator Wiring

the thermostat would turn on the actuator so flow through the circuit could resume.

Some valve actuators are supplied with two wires. These are the minimum wires needed to operate the actuator. Other valve actuators are supplied with four wires. Again two are needed to operate the actuator. The other two are connected to an isolated end switch inside the actuator. This switch closes whenever the actuator reaches its fully open position. They are typically used to signal the heat source and distribution system that heat is required by the zone associated with this actuator. The heat source is operated if necessary, as is the distribution circulator.

Other Manifold Accessories:

Some manifolds also have the option of being equipped with individual flow meters for each circuit. The flow meters can be used in conjunction with the supply and return temperature of a given floor circuit to confirm the heat output of the circuit.

Some manifold stations are also supplied with purging valves that allow hoses to be connected to both the supply and return manifolds for filling and purging. Pressurized water from the hose is forced into the supply manifold at a high flow rate. As it moves through the floor circuits, it pushes most of the air in the tubing toward the return manifold and eventually out the other hose.

Isolation valves are another option offered on some manifolds. When present, they allow a given manifold station to be isolated from the remaining system piping. This is very helpful during the previously described purging process, or if maintenance/repair operations need to be done. Typically these valves are either ball valves or butterfly valves that induce minimal pressure drop in their normal, fully open position. For manifolds not supplied with such valves, a full port ball valve can be piped in-line with each manifold during installation.

Air venting devices are supplied with some manifold assemblies. They can be either a manual venting valve, or a float-type automatic air vent. In either case, the air vent is usually located on the return manifold. This allows purging flow to push air toward the vent.

Manifold Enclosures:

All manifold stations should be placed where they're easily accessed during system start up and if service is required. In some cases manifolds are located in the mechanical room. Here there is typically no need to conceal them. In other cases manifold stations are located within the finished areas of the building. Here attention must be given to some type of enclosure.

Some manufacturers provide ready-to-install steel enclosure cabinets complete with hinged doors and knockouts for tubing and wiring. Depending on the length of their manifolds, some enclosures will fit between standard 16" or 24" o.c. wall framing. In other cases the framing must be adjusted to accommodate the enclosure.

Another option is to build the enclosure for each manifold. Many options exist depending on the type of wall finish and budget for the project. Details for one of the simplest enclosures suitable for wood frame walls is shown in Figure 4-9. The manifold brackets are fastened to a plywood plate that's "set in" between the studs so as not to interfere with drywall (or other

wall finish). After the drywall is in place, wood casing is used to frame the opening. A lip of about 1" width is left inside the casing. It provides a stop for a plywood panel to rest against. The closure panel can be painted, covered with wall paper, or otherwise finished as desired. The panel can be held in place by magnetic latches like those used for kitchen cabinet doors. Another option is to build a hinged cabinet-

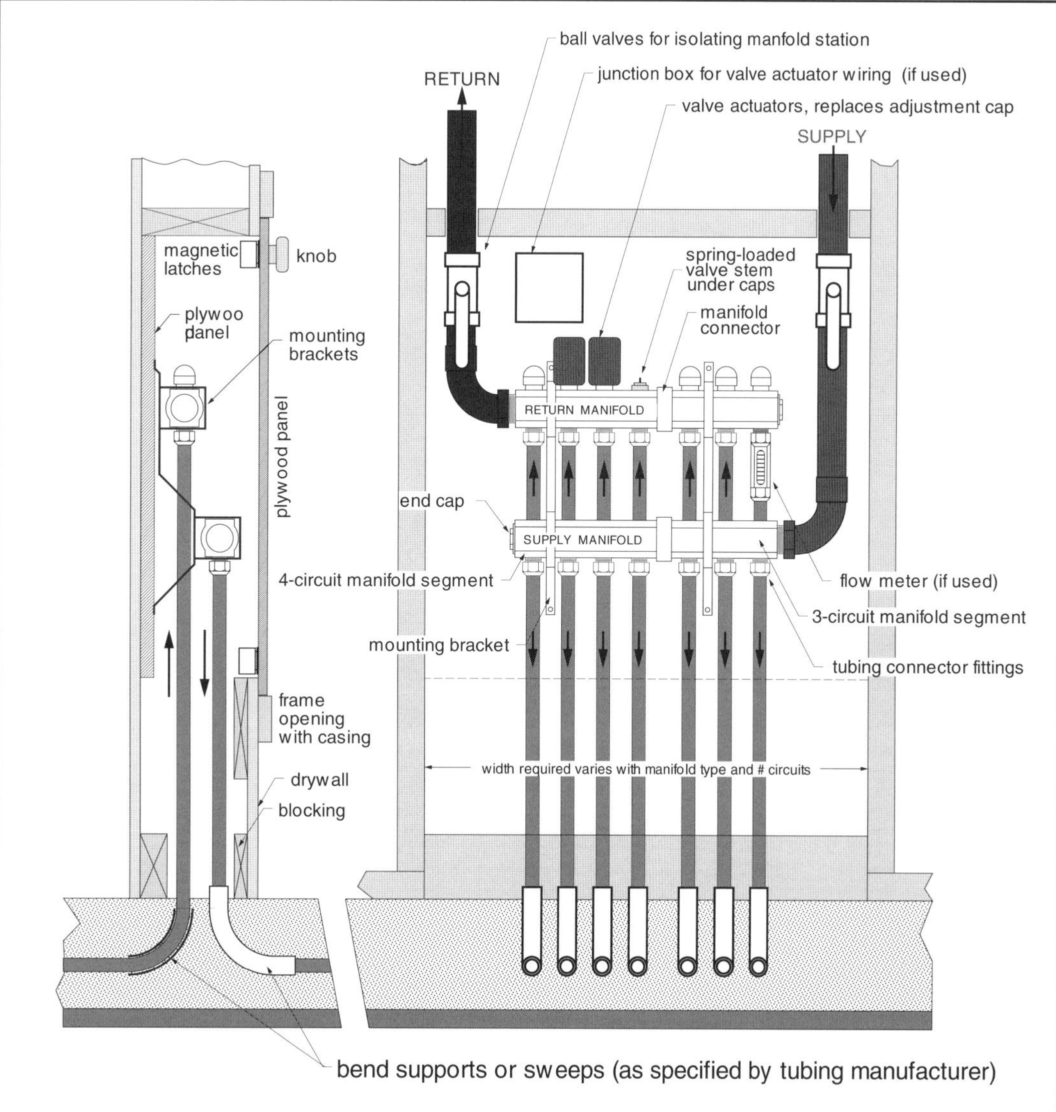

- The details shown are representative only. Other configurations are possible. Coordinate details and opening sizes with building designer and general contractor.
- 2x6 stud cavity, or furred out 2x4 cavity of 5" depth is adequate for most manifolds
- Verify exact manifold placement with building plans to ensure manifold stays within stud cavity.

Figure 4-9 Representative Manifold Station for Floor Heating System

style door, but this obviously adds cost to the job. Coordinate with the building designer and general contractor regarding the placement, width, height, and finish details for all manifold enclosures.

4•4 Tubing Installation Procedures

This section discusses the installation of tubing for slab and thin slab radiant panel systems which comprise a significant portion of hydronic radiant floor installations. A gallery of photographs of some of the other methods is also included. Keep in mind that the sequences presented are generic and reflect experience with existing products. New products, as well as new installation techniques, are always being developed as the radiant panel industry grows. New products will likely change the market share of different installation methods Again it is prudent to obtain and follow manufacturer's installation instructions when they're available.

Slab-on-grade Systems:

The installation of hydronically heated slab-on-grade floors varies depending on the specific products being used, as well as the design of the slab and its reinforcing. A typical sequence for a residential/light commercial slab using welded wire reinforcing is as follows:

1. Verify that all underslab plumbing and electrical services are in place.
2. Finish the grading by leveling and tamping any fill. The final grade under the slab should be accurately leveled allowing for the thickness of both the slab and under-slab insulation. Any loose rocks should be raked off providing a smooth stable surface.

Figure 4-10 Preparing Concrete Base

3. Cover the subgrade with a minimum 6-mill polyethylene moisture barrier. All edges should be well overlapped. This barrier prevents soil moisture from migrating upward and possibly affecting the finish floor materials or adhesives.
4. Place the underslab and slab edge insulation. A few spots of construction adhesive may be needed to secure edge insulation in place until

Figure 4-11 Place Insulation

backfilling. It's also advisable to weight down the foam boards as they are placed to prevent wind uplift. In many jobs this can be done by installing the welded wire reinforcing as soon as the foam boards are placed.

5. Complete the installation of all weld wire reinforcing. Adjacent sheets should be overlapped and tied.
6. Tubing placement is greatly simplified by preparing an accurate tubing layout drawing ahead of time. This drawing should be accurately scaled and show the location of all walls, partitions, tubing circuits, and manifold locations. An example of such a drawing is shown in Figure 4-5 .
7. Carefully locate the manifold station(s). Often times they are to be installed within interior walls, and thus must be accurately placed to ensure they will be concealed when the walls are installed. Although techniques differ, many installers choose to erect a temporary frame that supports both a supply and return manifold at each manifold location.
8. Before laying down pipe, it's helpful to mark out portions of each circuit directly on the underside insulation using spray paint. Locate and mark return bends, corners, or other places where the circuit changes direction.
9. Begin each circuit by selecting a coil of tubing that's at least as long as the circuit to be placed. Fasten one end of the coil to the supply manifold and proceed to uncoil the tubing from its roll. Never pull the tubing off the side of the roll. As it's unrolled, the tubing should be fastened to the welded wire reinforcing every 2 to 3 feet on straight runs, and closer as necessary to hold down any return bends or

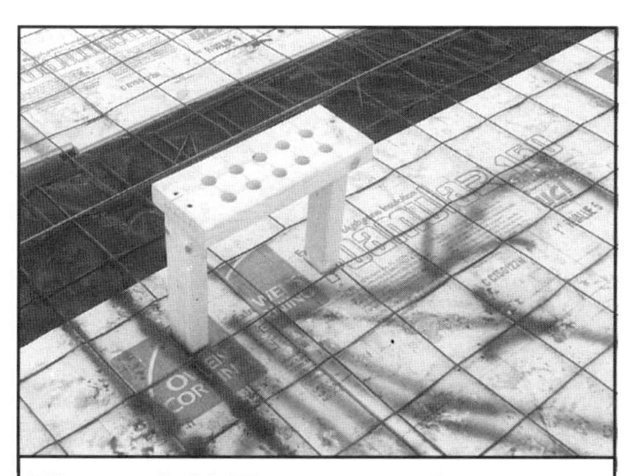

Figure 4-12 Temporary Support

Figure 4-13 Connect Tube to Manifold

Figure 4-14 Tie Tube to Mesh

other shapes created. Complete the circuit by cutting the tubing and fastening its end into the return manifold. Be sure the tubing is long enough before it's cut. Follow manufacturer's recommendations for detailing the locations where the tubing will eventually penetrate the top of the slab. Be careful not to kink the tubing where it bends from horizontal to vertical under the manifold stations. Whenever possible avoid crossing one tube over another. Again careful planning is the key to neat and accurate tubing placement.

10. After all circuits are placed they must be pressure tested. Follow the manufacturer's recommendations whenever available. The RPA Standard Guidelines specify a pressure test of 1.5 times the operating pressure, or 100 psi (whichever is greater) for a minimum of 30 minutes. Verify that no leaks are present.
11. If the slab will have control joints, any tubing that crosses through (or under) a control joint should be sleeved. Usually a sleeve consists of a piece of polyethylene or PVC pipe, about 1 foot long, that loosely fits over the tubing, and is centered under the control joint. Split sleeves are easy to install after the tubing is in place.
12. The concrete can be placed several ways. Use of an overhead boom and pump system is preferably when available since it minimizes traffic over the tubing. As the concrete is placed the tubing may be lifted to approximately half the thickness of the slab. This is generally done using steel hooks to lift the welded wire reinforcing. Do not hook and lift the tubing directly. Do NOT lift the tubing in the vicinity of control joints. Follow

Figure 4-15 Pressure Test Tubing

Figure 4-16 Placing Concrete

manufacturer's recommendations on how much pressure (if any) should be present in the tubing during the pour. The concrete is finished in the normal manner.

Concrete Thin-Slabs:

Thin-slab floor heating systems can also be constructed of portland cement-based concrete. One concrete recipe that has been used for heated thin-slabs is given in Figure 4-17. Admixtures such as a superplasticizer, water reducing agent, and fiberglass reinforcing give the concrete the "flowability" to encase the tubing, as well as low shrinkage characteristics that minimize cracking. This formulation of concrete weighs about 140 pounds per cubic foot. This adds approximately 12 pounds per square foot to the floor's dead loading per inch of slab thickness. For example, a typical 1.5" thick concrete slab adds about 18 pounds per square foot to the floor's dead-loading. The same floor height considerations mentioned for gypsum-based thin-slabs also apply to concrete thin-slabs.

One installation method calls for the thin-slab to be constructed before any exterior walls or interior partitions are erected. The installation sequence goes as follows:

1. Prepare the subfloor by sweeping clean.
2. Use a chalk line to mark the location of all walls and partitions directly on the subfloor. Also mark the location of base cabinets, and

MIX Design for 1 cubic yard of 3000 psi @ 28 day concrete topping

Type 1 Portland cement:		517 lbs.
Concrete sand:		1639 lbs.
#1A (1/4" maximum) peastone:	1485 lbs..	
Air entrainment agent:	4.14 oz.	
Hycol (water reducing agent):		15.5 oz.
Fiber mesh:		1.5 lbs.
Superplasticizer (WRDA-19):		51.7 oz.
Water:		about 20 gal.

NOTES:

• Request pricing on this mix from local concrete batch plant.

• Coverage is approximately 200 square feet of floor per cubic yard at 1.5" slab thickness.

• This mix adds approximately 18 pounds per square foot to the "dead loading" of the floor it is installed on.

• Place this mix over polyethylene bond breaker sheet to ensure no bonding occurs with plywood substrate.

• Water reducing agent and superplasticizer give mix high slump (approx. 6") to for good "flowability" around tubing fastened to deck.

• Fibermesh increases tensile strength of mix to reduce shrinkage cracks.

• Use control joints every 8 to 10 feet to control shrinkage crack locations.

• Water content of mix may vary depending on dampness of stone and sand.

• Do not use frozen aggregate when preparing this mix. It could lead to excess water and resulting shrinkage.

• Do not place this mix in direct sunlight to avoid water evaporation from surface to hydration of cement.

Figure 4-17 Thin-Slab Concrete Mix

closet flanges for toilets.

3. Cover the entire area that will receive the thin-slab with clear 6-mill polyethylene. This provides a bond-breaker layer between the concrete and the wood subfloor. By not allowing the concrete to bond to the subflooring, the concrete is less likely to develop random cracking due to shrinkage in the subflooring. The polyethylene sheet also minimizes moisture absorption by the subfloor when the concrete is poured.

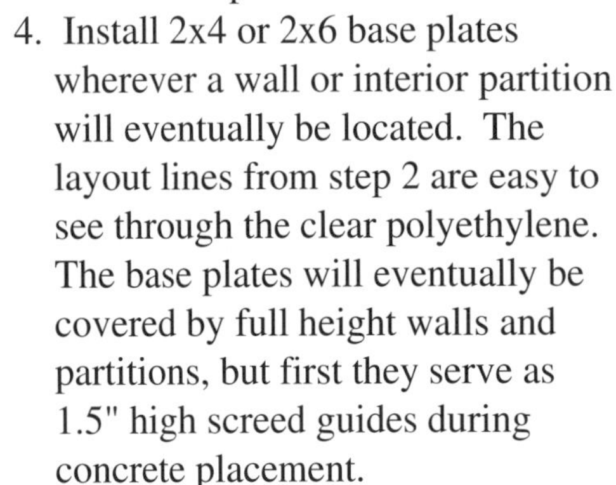

4. Install 2x4 or 2x6 base plates wherever a wall or interior partition will eventually be located. The layout lines from step 2 are easy to see through the clear polyethylene. The base plates will eventually be covered by full height walls and partitions, but first they serve as 1.5" high screed guides during concrete placement.
5. Install and pressure test all tubing circuits as described for gypsum thin-slabs.

Figure 4-18 Install Base Plates

6. Install control joint strips to divide the thin-slab into a "mosaic" of smaller areas. This helps prevent random cracking of the slab, and instead forces the cracks to occur along the control joint lines. Control joint strips should be installed anywhere the concrete is likely to crack. This would include doorways, inside corners in the slab, and any other "bottlenecks" in the slab. Larger floor areas should be broken into a grid pattern with control joints. Control joint spacings of 8 to 10 feet are generally adequate for thin slabs. Control joints should also be placed where transitions in finish flooring will occur.

Figure 4-19 Install Control Joints

All edges of the 2x base plates that eventually contact the concrete slab should be coated with a suitable release agent to prevent bonding. Mineral oil can be used, as can commercially available concrete form release agents. Be sure any release agent used is chemically compatible with the tubing used.

7. The final step is to pour the concrete. It can be placed from a wheelbarrow and carefully raked in place over the tubes. Take care not to place the nose bar of the wheelbarrow directly on the tubing when dumping the load. The concrete is

Figure 4-20 Place Concrete

screeded using the 2x base plates as height guides. On large areas it may be necessary to temporarily fasten a 2x4 to act as a height guide. After screeding the concrete is floated and finished in the usual manner. Wall framing can begin the next day. Narrow cracks should appear in the slab directly above the control joints after a few days of curing.

8. Before any finish flooring materials are applied, the thin-slab must be tested for a suitably low moisture content. A two foot square piece of clear polyethylene film is taped to the slab along all four of its edges. If, after a minimum of 48 hours has elapsed and there is no visible moisture under the film, the slab is generally considered dry enough for finish floor application. Note that drying times will vary considerably with location, season, interior temperature, and so forth. Never attempt to install finish flooring until the slab is adequately dried. Also be sure to follow any specific installation procedures specified by the flooring manufacturer. Cracks will form above the control joint strips within a few days after placing the concrete.

Figure 4-21 Raking and Screeding

Figure 4-22 Crack at Control Joint

Gypsum Thin-Slabs:

When gypsum-based thin-slabs are used, specific installation instructions provided by the manufacturer should be closely followed.

Gypsum thin-slabs are usually installed after the walls have been closed in with drywall or other finish materials. The highly flowable gypsum mix fills in any gaps between the drywall and the subflooring reducing air leakage and sound

transmission under walls.

A typical installation sequence is as follows:

1. Fasten all tubing to the subflooring using pneumatic staples, plastic clips, or other fasteners acceptable to the tubing manufacturer. These fasteners must not nick or scratch the tubing surface. The tubing should be fastened at intervals of 24 to 30 inches on straight runs, and at least twice at each return bend. All portions of the tubing circuits should be held tightly against the floor to prevent it from rising due to buoyancy when the gypsum underlayment is poured.
2. All tubing circuits must be pressure tested prior to the pour. The RPA Standard Guidelines specify a pressure test of 1.5 times the operating pressure, or 100 psi (whichever is greater) for a minimum of 30 minutes. Verify that no leaks are present.
3. Most gypsum underlayment systems begin with a sealant/bonding agent that is sprayed on the floor and tubing. This limits water absorption into the subfloor and enhances the bond of the gypsum underlayment to the subfloor.
4. The gypsum underlayment is prepared in a portable mixing rig located outside the building. Once mixed it is pumped into the building through a hose.
5. The gypsum underlayment is typically applied in two layers (called "lifts"). The first lift is poured equal to the depth of the tubing. As it cures, it shrinks slightly in the vertical direction relative to the height of the tubing. Within a short time, typically 1.5 to 2 hours, this layer is hard enough to walk on. A second layer of the material is then applied over the entire area. The second layer should have a minimum thickness

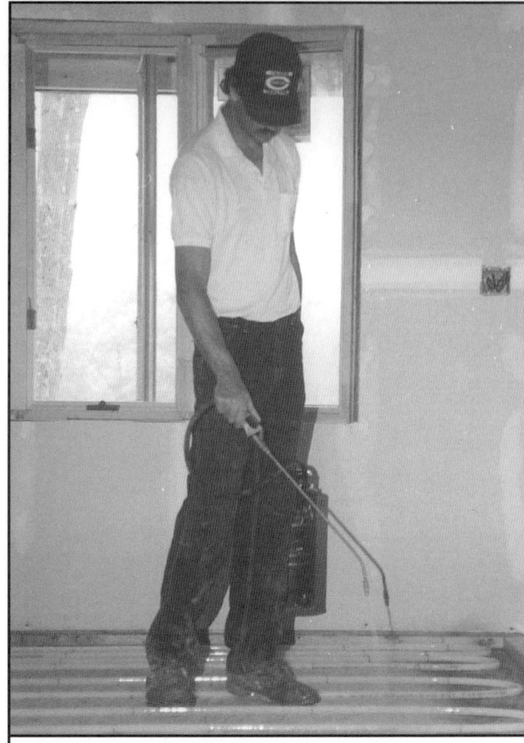

Figure 4-23 Applying Sealer

Figure 4-24 Gypsum Mixer/Pump

of 3/4" over the top of the tubing. It cures to a smooth flat surface hard enough to walk on in a few hours.

6. Before any finish flooring materials are applied to the thin-slab it must be tested for a suitably low moisture content. Again it is important to follow the manufacturer's recommendations. Typically, a two foot square piece of clear polyethylene film is taped to the slab along all four of its edges. If, after a minimum of 48 hours has elapsed and there is no visible moisture under the film, the slab is

Figure 4-25 Applying First Lift

Figure 4-26 Applying Second Lift

Figure 2-27 Raking and Smoothing to Depth

generally considered dry enough for finish floor application. Note that drying times vary considerably with location, season, interior temperature/humidity, and so forth. Never attempt to install finish flooring until the slab is adequately dried. Note also that certain primers/sealants may be required to achieve the proper bond between the thin-slab and various finish floors.

4•5 Pumps, Distribution Piping, and Controls

Thus far we've looked at different methods of creating hydronic radiant heating panels, as well as several options for generating the heat they need. To make these "sub-assemblies" work as a system, they must be linked by properly designed distribution piping, managed by suitable controls, and provided with properly sized circulators.

This section describes pumps, distribution piping and controls because they work together A choice of control method stipulates at least a portion of the distribution piping design and vice versa. A pump is always required. Pumps are often used in ways where they contribute to the control of the system.

Pumps:

The circulating pump is the heart of any hydronic heating system. Its job is to circulate heated fluid through the system pipes from the heat source to the heat emitters and back again. Pumps must be properly sized to accomplish this efficiently. Two factors determine selection of the pump; flow rate and pressure loss.

The flow rate is determined by the amount of heat required to be delivered and the ΔT of the water in the system. The amount of heat is the Btuh output determined from a heat loss analysis of the building. The ΔT is the difference in temperature of the supply water as it leaves the heating appliance and the return water after it has circulated through the system and delivered its heat into the dwelling. The flow rate is usually measured in Gallons Per Minute (GPM).

Pressure drop is the resistance to flow through the system caused by friction. Pressure loss is usually measured in feet of head or pounds per square inch (2.31 ft of head = 1 psi).

When both the flow rate and pressure drop are known, the pump can be selected from manufacturers' performance curves. The pump is selected so that the system operating point is near but within the pump performance curve as shown in Figure 4-28.

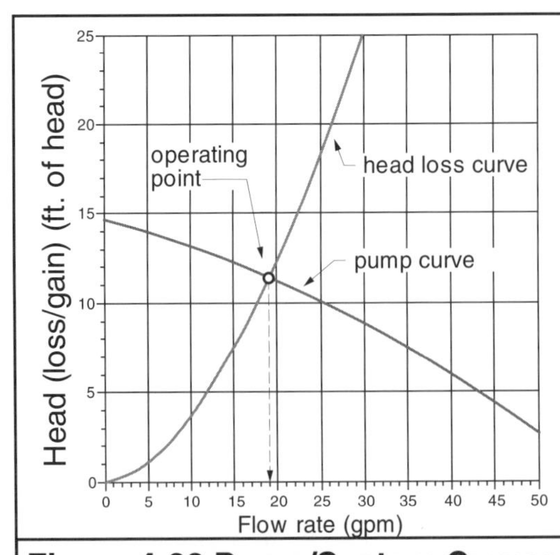

Figure 4-28 Pump/System Curves

Sometimes it is advantageous to obtain more flow or pressure by using two pumps. The advantage of this is that smaller pumps are more common and less expensive than larger ones. Two circulators installed in series will approximately double the pressure. Two circulators installed in parallel will approximately double the flow.

Reset Control:

Piping and control strategies can be broadly classified into two groups: those that operate the radiant panel at a fixed supply temperature when heat is required, and those that reset (e.g. adjust) the water temperature to the radiant panel based on outside temperature. There are several ways of implementing each of these control schemes.

Fixed temperature control means the water temperature supplied to the radiant panel when heating is required remains within a narrow range centered on a single "target" temperature selected by the system designer.

Reset control means the water temperature to the radiant panel varies as necessary to make the heat output of the panel equal the heat loss of the building.

It's important to understand the concept behind reset control before trying to incorporate it into a given system. Reset control is based on the premise that a building's heating load changes in proportion to the temperature difference between inside and outside air temperature. At the same time, heat output from a radiant panel system is approximately proportional to the temperature difference between inside air

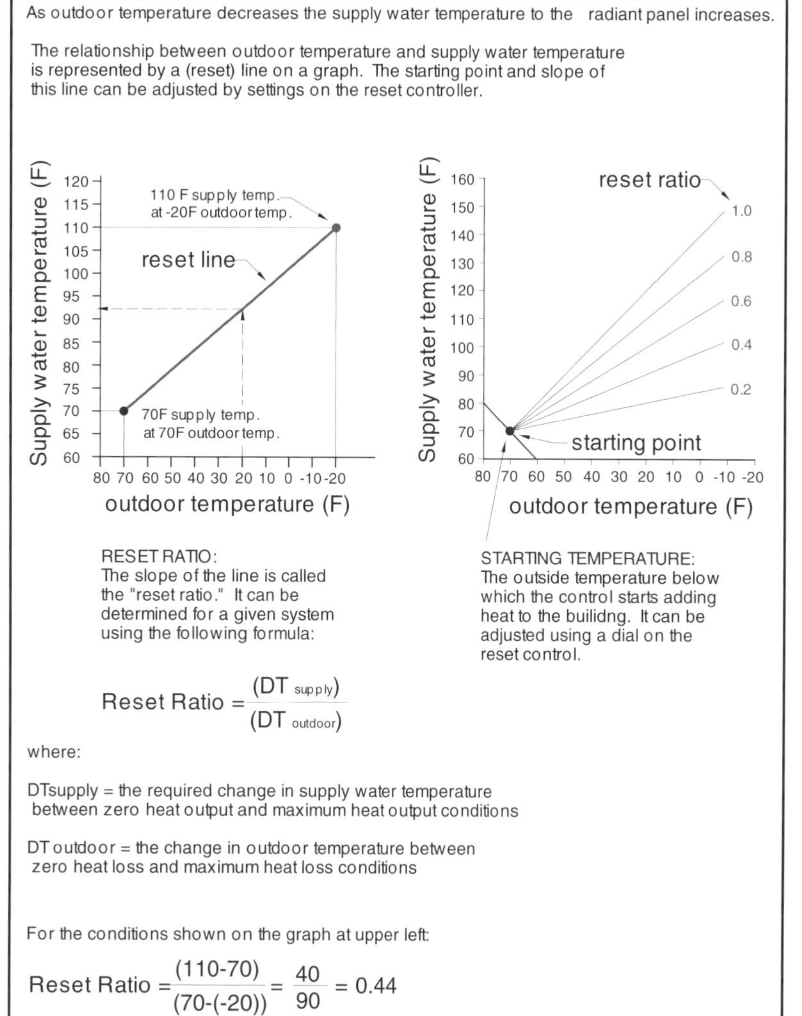

Figure 4-29 Reset Control Ratio

temperature and the water temperature in the panel. The mathematics of these two relationships "imply" a third proportionality between water temperature in the panel and outdoor temperature. This relationship forms the basis of reset control.

The graph in the upper left of Figure 4-29 shows a typical relationship between water temperature and outdoor air temperature for a radiant panel system requiring 110 °F water to match the building's heating load when it's -20 °F outside. As the outdoor \temperature climbs, the "target" (ideal) water temperature supplied to the radiant panel to meet the building's heating load decreases along the straight line. For example, when the outside temperature is 20 °F, the ideal water temperature in this particular radiant panel system is about 92 °F When outdoor temperature finally climbs up to 70 °F (and assuming the desired temperature in the building is also 70 °F), the water temperature supplied to the radiant panel is also 70 °F, and thus heat output from the radiant panel is zero.

The slope of the line in this graph is called the reset ratio, and is set using a dial on the control. To determine what reset ratio is needed for a given radiant panel system, you need to know what water temperature the panel requires under design load conditions, as well as the outdoor design temperature. You then put these numbers into the formula in Figure 4-29.

The graph in the upper right of Figure 4-29 shows that varying the reset ratio changes the slope of the line . Most low temperature radiant panel systems will have a reset ratio in the range of 0.3 to 0.8. Medium and higher temperature radiant systems, such as those using metal heat transfer plates or suspended tube applications, typically require higher reset ratios in the range of 1.0 to 2.0.

Most reset controls also allow the starting temperature of the line to be adjusted. It can be adjusted upward if the radiant panel requires higher supplying water temperatures during mild weather. Or it can be adjusted downward to compensate for internal heat gains in some building. Some reset controls, when equipped with an indoor temperature sensor, can even make these adjustments automatically. Since adjustment strategies vary from one control to another, it's best to consult the manufacturer's manuals for details on properly setting a reset controller.

Water Temperature Control:

Many radiant panel systems operate at relatively low water temperature in comparison to other types of hydronic heating systems. That's because the large surface area of the radiant panel (be it a floor, ceiling, or wall) can release all the heat necessary to maintain comfort without climbing to a high temperature. In many cases, the system's heat source is capable of generating water temperatures higher than needed by the radiant panels, especially during mild weather. In such cases a water temperature control strategy is needed to prevent the panels from overheating the space.

This can be done in a number of ways, such as cycling the heat source on and off, or using one of several piping schemes that mixes hot water from the heat source with cooler water returning from the radiant panel circuits. This section surveys most of the currently used methods of water temperature control.

Direct-Piped Systems:

One of the simplest piping/control strategies is where the heat source and radiant panel are directly connected to form a series piping circuit as shown in Figure 4-30. This configuration is suitable for heat sources that can operate at the relatively low water temperatures required by most radiant panels without being adversely effected by flue gas condensation. The following heat sources make this approach viable:

- Tank-type water heaters
- Condensing boilers with low flow resistance
- Thermal storage tanks such as those used in solar, off-peak electric, and some solid fuel boilers
- Electric boilers

Cycling the heat source on and off controls the temperature of the supply water in a direct-piped system. When the temperature of water leaving the heat source climbs slightly above the temperature needed by the radiant panel, a "high-limit control" temporarily stops the heat source from producing more heat. The distribution circulator, however, continues to run.

The radiant panel now dissipates heat stored in its thermal mass, as well as the thermal mass in the heat source and distribution piping. Since heat is being released by the radiant panel, but not being replaced by the heat source, the temperature of the system water decreases. The greater the thermal mass of the system, the slower the temperature drops.

Eventually the heat source must be turned on if the radiant panel is to maintain comfort in the building. This cycling operation of the heat source continues as long as the thermostat in the space continues to call for heat. When the thermostat is "satisfied" (e.g. that no further heat input is required) the heat source is turned off (regardless of water temperature). The distribution circulator between the heat source and radiant panel may or may not be turned off.

The latter approach, known as continuous circulation, is usually preferred, especially with high thermal mass systems. It tends to reduce temperature swings in the heated space and "purges" the residual heat from the heat source and distribution piping into the heated space.

Several precautions must be observed in direct-piped radiant panel systems. Since the radiant panel circuits and heat source

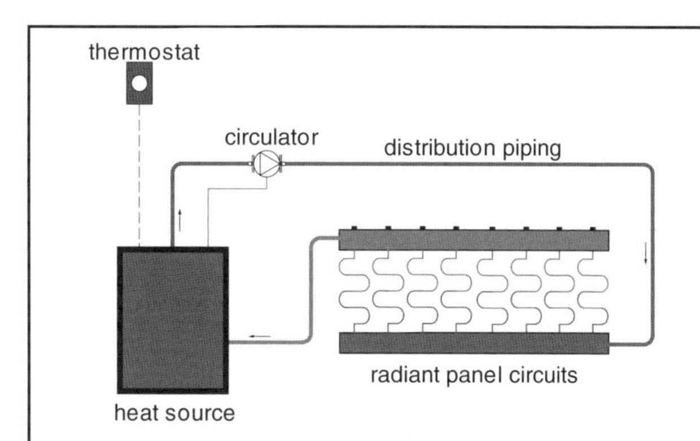

NOTE:

- Water temperature is controlled by switching the heat source on and off. Either fixed temperature or reset control can be used.
- Pressure drop of heat source, distribution piping, and radiant panel circuits add together. A high head circulator may be required in some systems.
- Water returning to the heat source is at the same temperature as water leaving the radiant panel circuits.
- **This arrangement is not recommended for conventional boilers because it can cause sustained flue gas condensation.**

Figure 4-30 Direct-Piped System

form a series circuit, the flow rate through the heat source will always be the same as that through the manifold(s) supplying the radiant panel circuits.

If, for example, the heat source requires a specified minimum flow rate that is greater than the flow rate through the distribution system, the heat source will be "starved" for flow. This is especially relevant to hydronic heat pump systems and low mass, copper tube boilers. Low flow through either can lead to problems. In the case of the heat pump, automatic shut down due to high refrigerant pressure is likely. In low mass boilers, problems such as steam formation with ensuing expansion noises can occur. This flow starvation situation also limits circuit-by-circuit zoning in direct-piped systems using heat sources requiring minimum flows while operating.

In direct-piped systems the head loss of the radiant panel circuits, distribution piping, and heat source are additive. Their total head loss must be overcome by the system's circulator. When the head loss of the heat source is small (storage tank type heaters, for example) this situation is not necessarily a problem. But other heat sources like hydronic heat pumps and some low mass boilers often have high head loss characteristics due to their internal design.

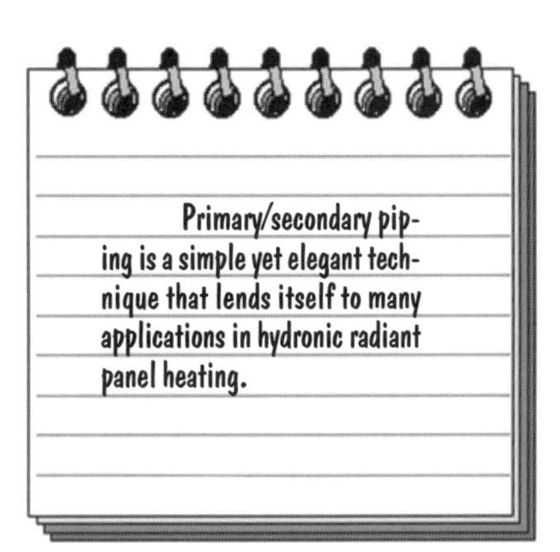

Putting a high head loss heat source in series with radiant panel circuits (that also tend to have high head loss) can result in a very high head requirement for the system's circulator. This is not to suggest that such high head losses cannot be overcome by high head circulators, or multiple circulators in series, but the economics of installing and operating high head circulator(s) should be carefully evaluated before making a final decision.

Direct-piped systems can use either fixed temperature control or reset control to turn the heat source on and off. When reset control is used, the high limit control on the heat source should be turned up a few degrees higher than the design water temperature of the system. This makes it function as a redundant safety control rather than the main operating control. Its function is to shut off the heat source only if the reset control allows the temperature to get too high. The reset controller is the primary operating control. Regardless of whether fixed or reset water temperature control is used, indoor temperature should be monitored by either an indoor temperature sensor or zone thermostats. This prevents overheating from solar heat gains, fireplace operation, or other internal heat sources.

By modifying the piping arrangement as shown in Figure 4-31 it's possible to use heat sources with high flow resistance in combination with fixed or reset water temperature control. This design uses the concept of "primary/secondary" piping to uncouple the flow resistance of the heat source from that of the distribution circuit.

Flow in the heat source circuit is generated by the primary circulator (P1). Many high flow resistance boilers come with such a circulator. In the case of a hydronic heat pump, an external circulator would be added to the primary circuit. The function of the primary circulator is to move water through the heat source and deliver it to the distribution circuit.

Flow in the distribution circuit is com-

pletely sustained by the circulator (P2). The interface between the primary and distribution circuits consists of two closely spaced tees. These allow heat to be delivered from the heat source to the distribution circuit, but at the same time prevent interference between the two circulators. This piping arrangement also allows heat sources with minimum flow rate requirements to be interfaced with a zoned distribution system in which flows vary over a wide range.

The piping shown in Figure 4-31 does NOT guarantee that the heat source is protected from low temperature return water. Assuming the heat source is properly sized, and that all zones of the radiant panel system are operating (e.g. the system is at design conditions), the water temperature entering the boiler will be about the same as the temperature leaving the return manifold. However, as the heating load decreases and zones begin to turn off, flow in the distribution system decreases. This increases the bypass flow across the primary/secondary tee junction causing the temperature in the boiler circuit to rise. Eventually the heat source reaches its high limit setting and shuts off.

In the case of condensing boilers the lower the high limit control is set, the greater the amount of flue gas condensation, and hence the higher the boiler's efficiency. However, setting the high limit control too low tends to cause frequent but short on-cycles. Some reset controls compensate for this effect by automatically widening the temperature differential the heat source operates at as the outside temperature increases.

thermostat
flow can go in either direction between the tees depending on the flow rates in both circuits
heat source circuit
P2
distribution circuit
closely spaced tees
P1
high limit control
radiant panel circuits
heat source
(with high flow resistance)

NOTE:

- Flow and head loss of heat source is handled by circulator (P1). Flow and head loss of distribution circuit and radiant panel circuits is handled by circulator (P2).
- This design allows flow in the distribution system to be "independent" of flow through the heat source. Reduced flow caused by zoning of radiant panel circuits will not effect flow through heat source.
- The tees at the primary / secondary connection should be copper, and spaced as close as possible.
- Water returning to heat source is not "guaranteed" to be hot enough to prevent flue gas condensation. Some boilers provide internal protection against flue gas condensation.
- Heat source can be operated by either a fixed temperature or reset control.

Figure 4-31 Primary/Secondary Piping

Mixing Assemblies:

When the heat source must operate at temperatures higher than the radiant panel, some type of mixing assembly is needed to interface the boiler circuit and distribution circuit. This concept is shown in Figure 4-32. The mixing assembly's purpose is to "meter" hot water from the boiler circuit into the distribution circuit such that the radiant panel is supplied with the appropriate water temperature. The hot water allowed into the distribution circuit mixes with cooler water

returning from the radiant panel circuits. When the correct proportions of hot and cool water are mixed, the result is a stream of water at just the right temperature needed by the radiant panel circuits. It's basically the same concept we use to adjust our shower temperature.

In radiant panel systems supplied by conventional boilers, the mixing assembly provides a second vital function. It protects the boiler from sustained flue gas condensation by ensuring water entering the boiler is above the dewpoint of the exhaust gases.

In radiant panel systems using conventional boilers the mixing assembly serves two purposes:

1. It provides the radiant panel with the appropriate supply water temperature.
2. It protects the boiler from operating with sustained flue gas condensation.

Any assembly of piping, valves, and controls that supplies a radiant panel with the proper water temperature could be considered a mixing assembly. Dozens of different methods have been used over the years, some more successfully than others. We'll look at all of the established methods, examining the strengths and weaknesses of each, as well as describing the kind of system each is best suited for.

Nearly all radiant panel systems that require mixing controls use one of the following strategies:

- Manually-adjusted valves
- 3-way (non-electric) thermostatic valves
- 3-way motorized valves
- 4-way motorized valves
- Injection mixing with 2-way valves
- Injection mixing with variable-speed pumps

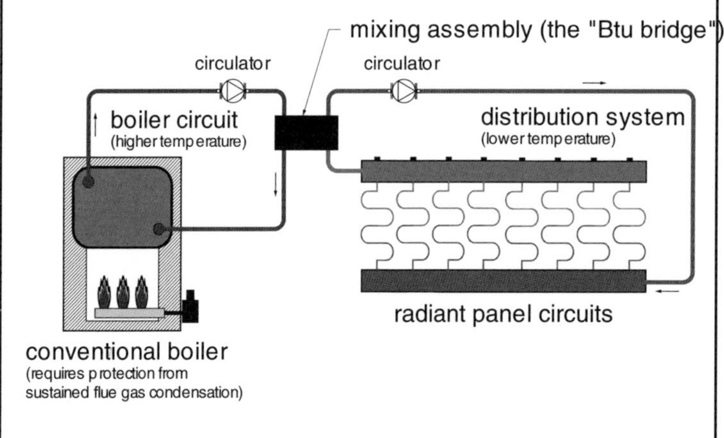

NOTE:

- The mixing assembly is the "Btu bridge" across which all heat entering the distribution system must pass. The temperature (and hence heat output) of the distribution system is controlled by regulating the heat flow across the mixing assembly.

- When used with conventional boilers, the mixing assembly must:

1. regulate water temperature supplied to radiant panel circuits.

2. Boosts water temperature entering boiler high enough to prevent sustained flue gas condensation.

Figure 4-32 Mixing Assembly Concept

Manually-Adjusted Valves:

The piping arrangements in Figure 4-33 shows how two manually adjusted valves might be used to control water temperature in the distribution circuit. The schematic at

the top allows lower temperature water returning from the radiant panel circuits to go directly into the heat source. This is fine for heat sources that don't require protection from flue gas condensation, but is seldom recommended for conventional boilers. The lower schematic includes a boiler circuit that bypasses some hot water between points (B) and (C) to boost the return temperature into the conventional boiler. Note, however, that this temperature boost will only be effective after the thermal mass of the system has warmed up to near normal operating temperatures.

In either system the goal is to set the two valves so that the right proportions of hot and cool water are mixed together at the tee labeled (A).

If the water temperature supplied by the boiler stayed constant, and the water temperature returning from the radiant panel circuits also stayed constant, it would be a simple matter to adjust the valves to achieve the desired (fixed) supply temperature. Unfortunately, these two temperatures change almost constantly.

For example, when the boiler first turns on, its water temperature is far lower than its normal operating temperature (which is usually close to its high limit setting). As the boiler water increases in temperature so would the temperature supplied to the distribution circuit. Likewise, if the radiant panel system has been off for a while, its thermal mass may be quite cool. This causes the return water temperature

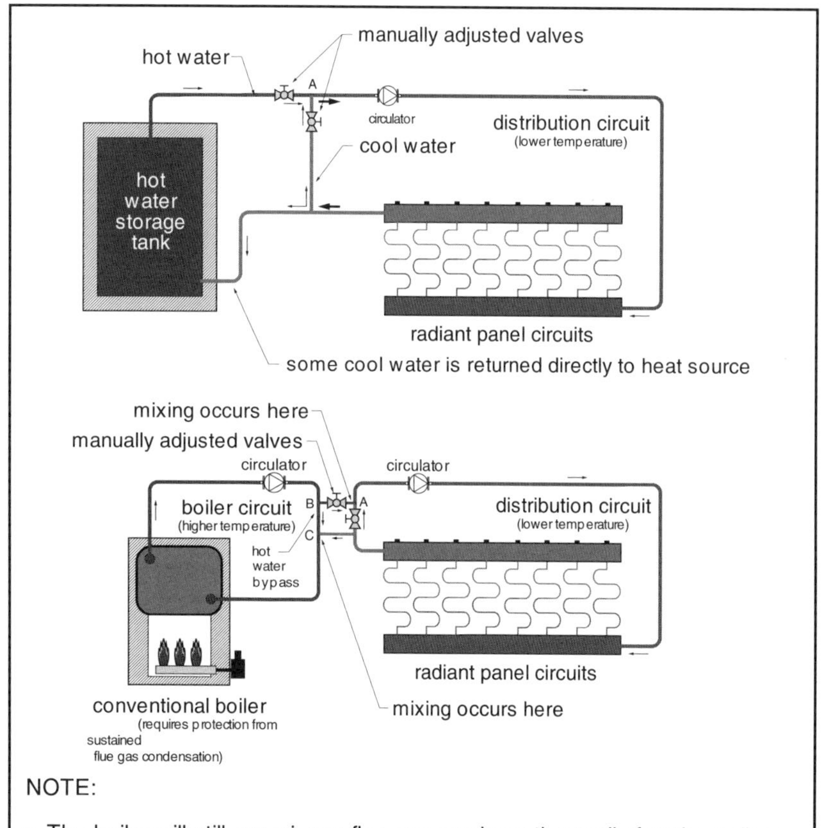

NOTE:

- The boiler will still experience flue gas condensation until after the radiant panel has warmed up. This could take several hours for a high thermal mass panel started from a cold temperature.
- The manually set valves are "blind" to water temperature. They cannot readjust flow proportions as the boiler warms and cools, nor as the radiant panel warms and cools. This can lead to wide variations in the water temperature supplied to the radiant panel and the boiler.
- This method of mixing is NOT recommended for use with high thermal mass radiant panels nor with conventional boilers.

Figure 4-33 Manually-Adjusted Mixing Valve Piping

to be lower than normal which also pulls down the supply temperature. If the radiant panel has high thermal mass, warm up will be very sluggish, in some cases taking several hours.

The basic problem with this mixing approach is that the valves cannot "react" to varying temperatures, and hence the supply water temperature they create will vary almost continually, in some cases over a wide range. Not only does this complicate the delivery of heat at the proper rate, it can even lead to temperature extremes that could physically damage part of the radiant panel as well as the heat source. Note also that any kind of manually-adjusted valve arrangement (be it a pair of 2-way valves as shown, a single 3-way valve, or even 4-way valve) would have similar problems. Because of their inability to adapt to changing water temperatures, manually-adjusted valves are really only suitable for systems that have essentially constant heat source and panel return temperatures, which unfortunately is almost never the case. Hence this method of water temperature control is seldom used in modern hydronic radiant panel systems.

3-way (Non-Electric) Thermostatic Valves:

The piping system shown in Figure 4-34 shows two ways of incorporating 3-way thermostatic valves as a mixing device. The internal construction of such valves allows them to automatically adjust the proportions of hot and cool water entering the valve. The knob on the end of the valve sets the thermostatic actuator to a selected "design" water temperature required by the radiant panel during peak load conditions. As the temperatures of both the boiler water and panel return water change, the thermostatic actuator moves the internal valve element in an attempt to maintain a constant mixed outlet temperature. If the temperature of the boiler water drops below the outlet temperature, the valve set for the cool port will be completely closed, and the valve will simply pass water entering from the boiler loop through to its outlet port. (See Figure 4-34.)

Most 3-way thermostatic valves have a specified temperature range over which they can operate. Different valves may be required depending on the temperature the radiant panel circuits must operate at. Be sure to check for the proper temperature range when ordering the valve.

Some 3-way thermostatic valves come with external temperature sensing bulbs connected to the valve actuator by a capillary tube. They operate essentially the same as valves with internal temperature sensing elements. It's very important to place the external temperature sensing bulb downstream of the mixing valve. The preferred location is downstream of the circulator in the distribution circuit. This ensures thorough mixing has occurred by the time the flow passes by the bulb. Improper placement of the sensing bulb can cause erratic operation. The bulb should be tightly strapped to the pipe and covered with pipe insulation to ensure accurate temperature sensing. Some valves are also supplied with fittings that allow the sensing bulb to be inserted into the flow stream at a tee. This method of mounting the bulb will yield faster valve response to changing temperatures.

The size of valve selected depends on the flow rate in the distribution circuit as

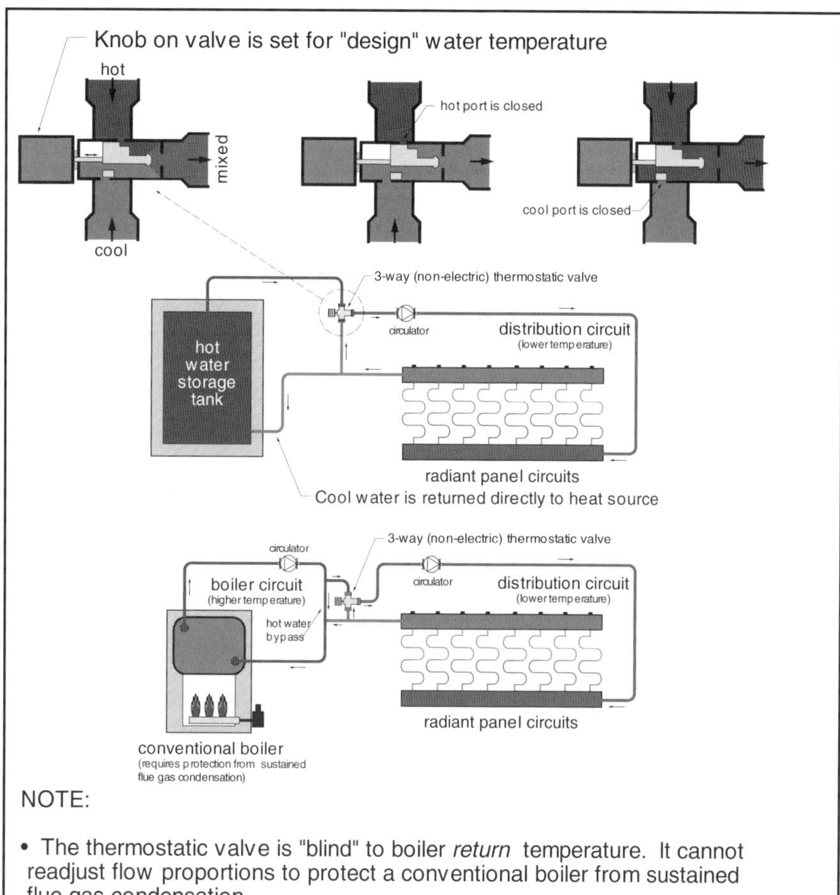

Figure 4-34 Thermostatic 3-Way Mixing Valve Piping

well as the valve's Cv rating. The Cv rating is the flow rate through the valve (in gallons per minute) at which it creates a 1 psi pressure drop. Generally speaking, the smaller the pipe size of the valve, the smaller its Cv rating. If the valve is undersized it will create a high head loss that the distribution circulator will have to overcome. The valve tends to "bottleneck" the distribution circuit. Undersized valves can also create objectionable noise if the flow velocity through them is too high. Some manufacturers have maximum recommended flow rates for their 3-way valves.

As a general rule, select a valve with a Cv rating approximately the same as the flow rate in the distribution system, and account for the valve's head loss when sizing the distribution circulator.

Three-way thermostatic valves supply the radiant panel circuits with a fixed water temperature regardless of the heating load. Under partial load conditions the radiant panel would overheat the building unless flow through the circuits is stopped when the heated space has attained the desired air temperature. There are a couple ways to accomplish this: When all the radiant panel circuits served by the 3-way valve are part of the same heating zone, a room thermostat in that zone can simply turn off the distribution circulator when the thermostat is satisfied. If there are several zones served by the same 3-way valve, individual valve actuators on the manifold are used to stop flow through their associated circuits as their thermostats require. The latter type of (room-by-room) zoning is often preferred when the building is divided up into several smaller rooms, each of which can experience different heat loads. A concept drawing showing this type of system is given in Figure 4-35, showing representative wiring as well as additional hydronic detailing.

3-way Motorized Valves:

The mixing ability of 3-way valves can also be combined with precise electronic control. The resulting motorized valve system can supply either fixed or reset temperatures to a radiant panel system. Figure 4-36 shows the concept.

The valve body used for this type of mixing system is different from that used for a 3-way thermostatic valve. It has a rotating shaft (in comparison to the linear motion shaft of a thermostatic valve). As the shaft is rotated through approximately

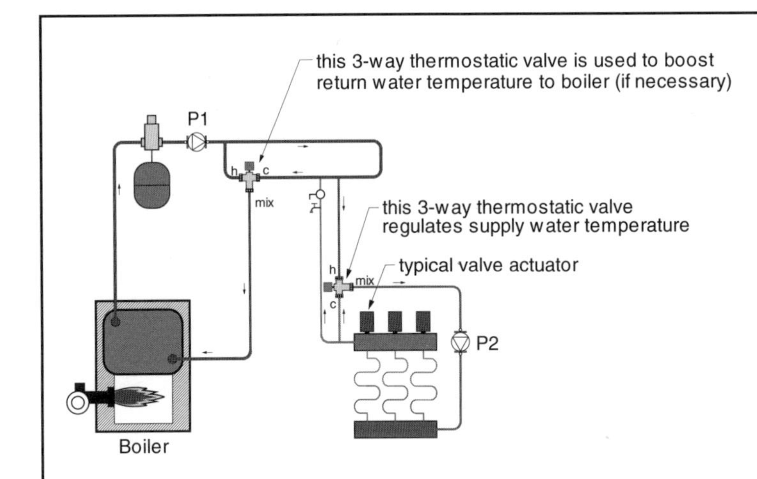

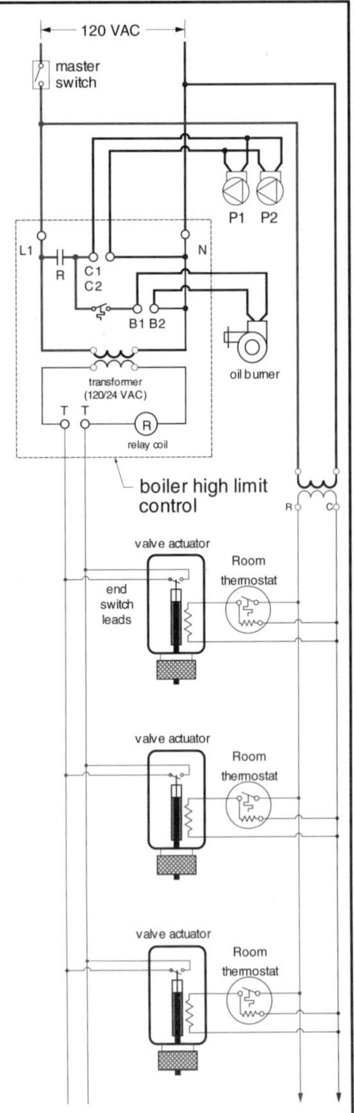

Operating sequence:

When a room thermostat calls for heat, 24 VAC is applied to the heating element of its associated valve actuator. In 3 to 5 minutes (for heat motor type actuators) the shaft of the actuator is fully retracted allowing the manifold valve to reach its open position. The actuator's end switch closes signaling the boiler high limit control that heat is requested. The boiler high limit control turns on the distribution circulator, and boiler circulator, and also enables boiler firing.

NOTE:

- The second 3-way thermostatic valve assured the boiler return temperature stays above the set minimum temperature (typically about 130 F).
- Because the boiler is protected against sustained flue gas condensation, this method of mixing is suitable for use with conventional boilers.

Figure 4-35 Three-Way Thermostatic Mixing with Individual Zoning

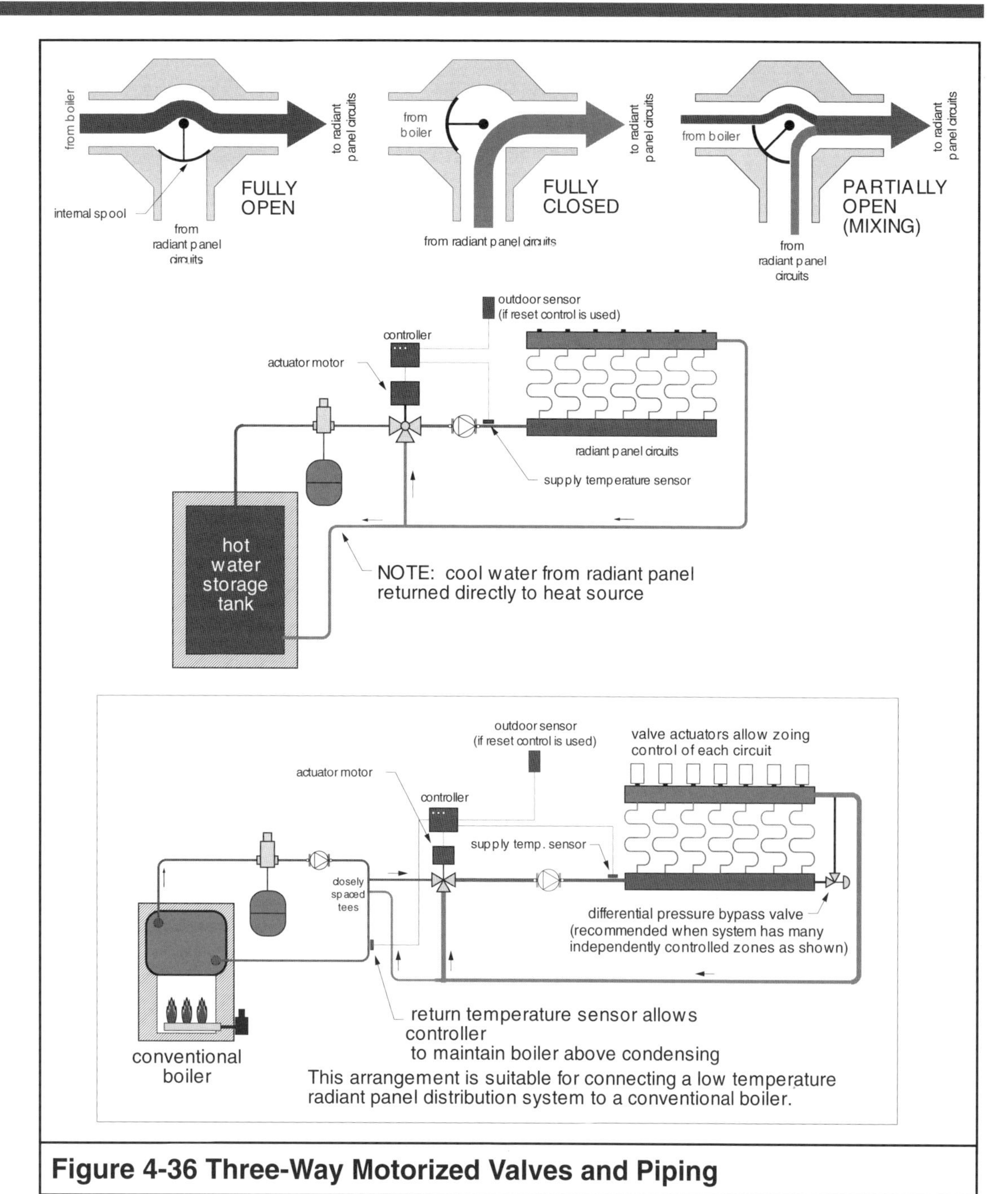

Figure 4-36 Three-Way Motorized Valves and Piping

90 degrees of arc, the internal spool simultaneously opens one inlet port and closes the other. This regulates the proportions of hot and cold water entering the valve, and hence adjusts its outlet temperature.

The actuating motor that rotates the valve spindle turns very slowly, but with considerable torque. Depending on the type of motor used (thermal versus geared), it can take 3 to 5 minutes for the shaft of the mixing valve to be rotated through 90 degrees. This slow rotation is not a problem given the slow response of many radiant panel systems. In fact, it helps stabi-

lize the system against overshooting and undershooting the desired supply water temperature.

A solid-state temperature sensor attached to the piping leading to the manifold constantly measures supply temperature. It provides feedback to an electronic controller that regulates the valve motor. If the temperature is exactly where it should be the motor does not change the valve stem's position. If the supply temperature is slightly low, the motor very slowly rotates the valve stem, allowing more hot water to enter the mix and thus increase outlet temperature. If the temperature is too low the motor rotates the shaft in the opposite direction, allowing more cool water into the mix. Since the sensor is downstream of the valve it's constantly providing feedback, allowing the mixing system to fine-tune supply water temperature. Under normal operating conditions such a mixing system can maintain extremely stable and accurate control of supply water temperature.

In addition to operating the mixing valve motor, many controllers can also sense the return temperature going to the heat source, and when necessary partially (or completely) close the mixing valve so this temperature stays above a minimum setting. Such controls are said to give "priority" to protecting the boiler from flue gas condensation. As boiler return temperature increases to a suitable value the control slowly reopens the valve, allowing hot water into the mixing valve. A return temperature sensor is usually recommended when 3-way motorized mixing valves are used with conventional boilers. This option is not required, however, when the heat source is not subject to damage from flue gas condensation.

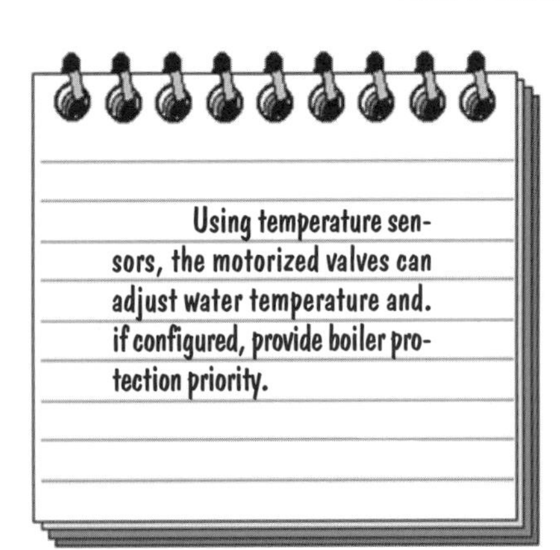

Most valve controllers can be set for either fixed temperature operation or outdoor reset control. In the latter case the reset ratio and starting point of the reset line can all be set on the control. Maximum supply water temperature (regardless of outside temperature) can also be set.

The piping design used with 3-way motorized valves is very similar to that used with 3-way thermostatic valves. When the heat source needs to be protected from flue gas condensation a boiler circuit with hot water bypass should be used.

On larger systems it's possible to reduce the size of the mixing valve by bypassing part of the return flow from the radiant panel circuits "ahead" of the valve as shown in Figure 4-36). In such systems the final mixing takes place in the tee downstream of the valve. It's crucial to install the temperature sensor downstream of this tee to sense the final mixed temperature.

4-way Motorized Valves:

The same actuating motor and electronic controller used with 3-way mixing valves can also be used with a 4-way valve. Such valves were developed specifically to provide mixing temperature control and return temperature protection for conventional (non-condensing) boilers. Although a 4-way mixing valve could be used with other types of hydronic heat sources, they usually don't require its dual functionality. The internal design of a 4-way mixing valve and its recommended piping are shown in Figure 4-37.

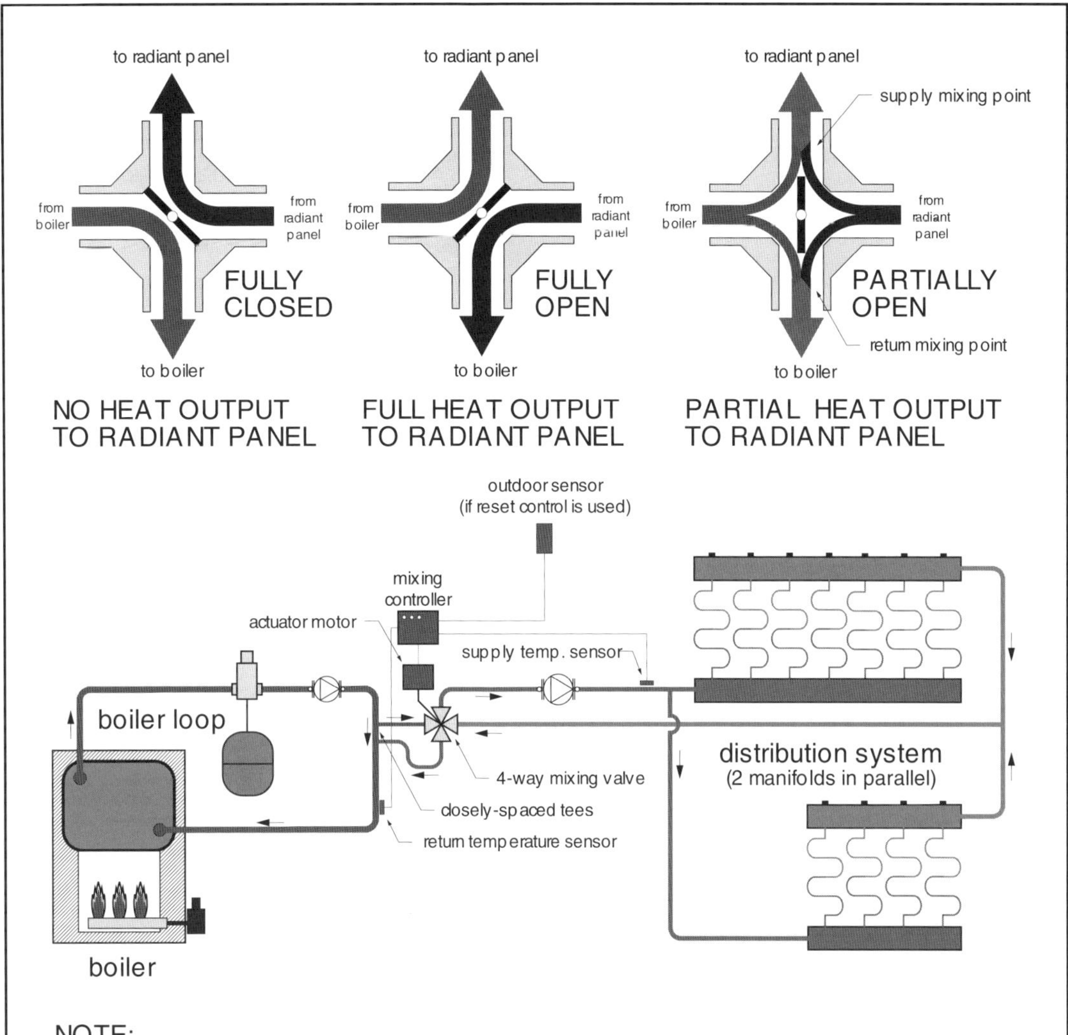

NOTE:

• Closely-spaced tees are used to connect boiler loop to 4-way valve. This eliminates interference between the boiler circulator and distribution circulator.

• Return temperature sensor allows mixing controller to maintain boiler return temperature high enough to prevent flue gas condensation.

• A 4-way mixing valve (body) without an actuating motor and controller can not provide consistent supply temperature to the distribution system. It also can not provide return temperature protection for the boiler.

Figure 4-37 Four-Way Motorized Valves and Piping

During normal operation hot water from the boiler is mixed with cool return water from the radiant panel circuits in two locations inside the 4-way valve. At the upper mixing point the hot and cool water streams are blended as necessary to attain the desired supply temperature to the radiant panel circuits. At the same time mixing also occurs in the lower valve chamber. Here the objective is to boost the temperature of the remaining flow of cool return water high enough that flue gas condensation will not occur when the mix is returned to the boiler. As with motorized 3-way valve systems, a temperature sensor mounted on the supply pipe to the radiant panel circuits provides feedback to the valve controller. Valve position is continually adjusted as necessary to maintain the proper supply water temperature to the radiant panel circuits.

The fact that 4-way motorized mixing valves can adjust themselves to provide stable supply temperature does NOT guarantee that return temperature to the boiler is always high enough to prevent condensation. For example, under cool-slab start-up conditions, when the 4-way valve is trying to boost supply temperature, it often needs all the hot water it can get from the boiler. In an attempt to get this hot water, the valve moves toward its fully open position.

Under these conditions there is simply not enough hot boiler water left over to boost the return temperature high enough to prevent condensation. In fact, when the 4-way valve reaches its fully open position all the cool water returning from the radiant panel is sent directly back to the boiler without any temperature boost. For this reason some controllers used with 4-way mixing valves have a return water temperature sensor that's mounted near the inlet to the boiler.

If this sensor detects low return water temperature the controller responds by partially (or completely) closing the mixing valve. This temporarily suspends heat output to the radiant panel allowing the boiler loop temperature to recover back to an acceptable temperature. Once boiler return temperatures are normal, the mixing valve is again allowed to open, sending heat to the radiant panel circuits. In effect, these controllers give "priority" to protecting the boiler. They allow a high mass radiant panel system to guide itself through transient (cool-slab) situations, unattended, and without creating conditions harmful to the boiler.

To avoid flow interference between the boiler circulator and distribution circulator, closely-spaced primary/secondary tees are used to connect the 4-way valve to the boiler loop as shown in Figure 4-37. In effect, these tees decouple the boiler loop from the distribution loop. The valve is still able to draw in hot water from the boiler loop because of the momentum effect created by return flow from the distribution loop. At most times there will also be a bypass flow of hot water through the primary/secondary tees that further assists in boosting boiler return temperature.

Manufacturers of 4-way valves should be consulted regarding valve sizing. In many cases (especially with low temperature radiant panel systems) it's not necessary to "line size" the 4-way valve to the pipe size of the distribution system. Doing so usually leads to purchasing a larger (more expensive) valve than required. It can also lead to the valve operating over a very limited travel range. This happens because the valve simply doesn't need to allow much hot water into the mix to achieve the necessary supply temperature, even under design load conditions.

Injection Mixing:

Another method of mixing hot boiler water with cool water returning from radiant panels (or other hydronic heat emitters) is called injection mixing. It's one of the simplest and most versatile concepts available for controlling water temperature in hydronic systems, and can be implemented in several different ways. It has especially favorable characteristics when used with low temperature radiant panel heating systems. The fundamental concept of injection mixing is shown in Figure 4-38.

Think of the distribution loop shown in this figure as a constantly circulating "conveyor belt" for carrying heat. When heat needs to be transported to the radiant panel, a small flow of hot water is pushed, (or "injected") into the distribution loop. The hot water mixes with cooler water returning from the radiant panel circuits at the tee labeled (mixing point). If the proportions of hot and cool water are just right, the mixed flow supplied to the radiant panel circuits is at the desired (target) value. Because the distribution system is completely filled with water, a portion of the cool return water must exit the distribution loop at the return injection riser tee. The flow rate of the exiting water will always be the same as the injection flow rate of hot water.

The temperature of the distribution loop, and hence its heat output, can be completely controlled by the rate of hot water injection. When heat is not needed by the radiant panel circuits served by the distribution system, no hot water should enter. The hot water injection rate must increase as the heating load of the panels increases. Any device that can regulate the injec-

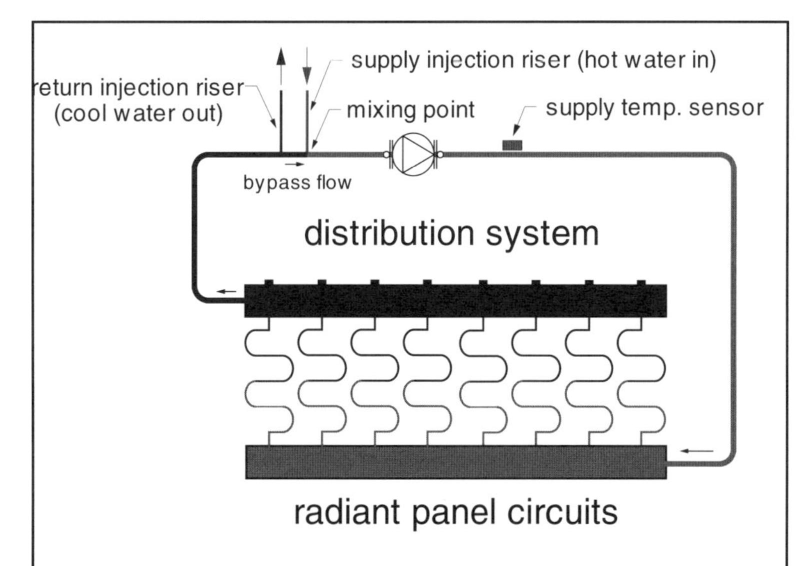

NOTE:

- The greater the incoming flow rate of hot water (through the suuply injection riser), the warmer the distribution system becomes, and the greater the heat output of the radiant panel circuits.
- The incoming flow rate of hot water is always balanced by an equal (exiting) flow rate of cool water.
- The preferred location of the supply temperature sensor is downstream of circulator. This ensures complete mixing has occure
- There are several strategies for controlling the rate of hot water injection.

Figure 4-38 Injection Mixing Concept

4

tion flow rate can thus regulate heat output from the distribution system.

The hot water injection flow rate needed to achieve a given rate of heat transfer to the distribution system can be calculated with the following equation:

$$f_i = \frac{Q}{490 \times (T_{hot} - T_{cool})}$$

Formula 4-1

where:

F_i = required hot water injection flow rate (gpm)

Q = rate of heat transfer to the distribution loop (Btuh)

T_{hot} = temperature of the hot (injection) water (F)

T_{cool} = temperature of water returning from the radiant panel circuits (F)

490= a constant based on the properties of water at an average temperature of 140 °F

(this value decreases to about 450 for 50% solutions of propylene glycol)

For example: What flow rate of hot (180 °F) boiler water is required to transfer 100,000 Btuh into the distribution loop of a radiant panel system that has a return water temperature of 95 F?

Solution: Putting the numbers into the formula and solving:

$$f_i = \frac{100{,}000}{490 \times (180 - 95)} = 2.4$$

It may be surprising how small the required injection flow rate is, especially considering this is a substantial heating load. This low flow rate could easily be carried through a 1/2" tube.

The "key" to understanding why such low flow rates can still transfer substantial amounts of heat is to look at the temperature difference between the entering hot water and the exiting cool water. For the system in the above example this difference is 180 - 95 = 85 °F, a very large value by comparison to temperature drop that occurs across many hydronic heat emitters (including radiant panel circuits). The fact that such a large temperature difference can exist at this point in the hydronic system allows the flow to be quite low, and still transport heat at the required rate.

A good analogy to this situation is the theory behind high voltage electrical transmission lines. By operating the conductors at a high voltage, the electrical current necessary to transfer a given amount of power is quite low. Since conductor size is determined by current rather than voltage, relatively small diameter wire can be used to reduce cost and weight. In the injection system the large ΔT between the boiler water and cool return water is analogous to the high voltage. It allows the flow rate (which is analogous to current) to be quite small for a given rate of heat transfer.

The lower the temperature of the return water from the radiant panel circuits, the lower the injection flow rate has to be. The opposite is also true. As the return water temperature increases, so must the injection flow rate in order to maintain a given rate of heat transfer (assuming the boiler outlet temperature stays the same). This is why injection mixing is optimally suited for low temperature radiant panel systems. Even at design load conditions, such systems have return water temperatures in the range of 90 to 110 °F However, not all radiant panel systems produce such low return temperatures, especially under design load conditions. For example, a plate-type radiant floor panel may have a

return water temperature in the range of 130 °F. If we assume the same 180 F injection water temperature and the same 100,000 Btuh load as used in the previous example, the required injection flow rate now becomes:

$$f_i = \frac{100{,}000}{490 \times (180 - 130)} = 4.1$$

This is substantially higher than the 2.4 gpm injection flow required for the low temperature radiant panel system of the previous example. It's easy (and interesting) to put other numbers into the formula to see the effect on the required injection flow rate. The following statement summarizes our discussion:

The greater the temperature difference between "hot" water from the heat source, and "cool" water returning from the radiant panel circuits, the smaller the injection flow rate will be for a given rate of heat transfer. Injection mixing thus reveals its greatest benefits when used in combination with a higher temperature heat source and lower temperature radiant panel.

Injection Mixing with 2-way Valves:

One method of applying injection mixing is shown in Figure 4-39. The higher temperature boiler loop is connected to the lower temperature distribution loop with two piping segments called injection risers. A 2-way thermostatic valve is installed in one of the risers. This valve is adjusted to the desired supply water temperature of the radiant panel. Its temperature sensing bulb is mounted on (or in) the supply piping to the radiant panel. If the supply water temperature at the bulb's location starts to drop, the valve responds by opening to allow more hot water into the distribution loop. Likewise, if the bulb detects that the distribution loop is climbing above the setpoint temperature the valve responds by reducing the hot water injection flow. When properly selected, such a valve can maintain a relatively steady supply temperature over a wide range of conditions.

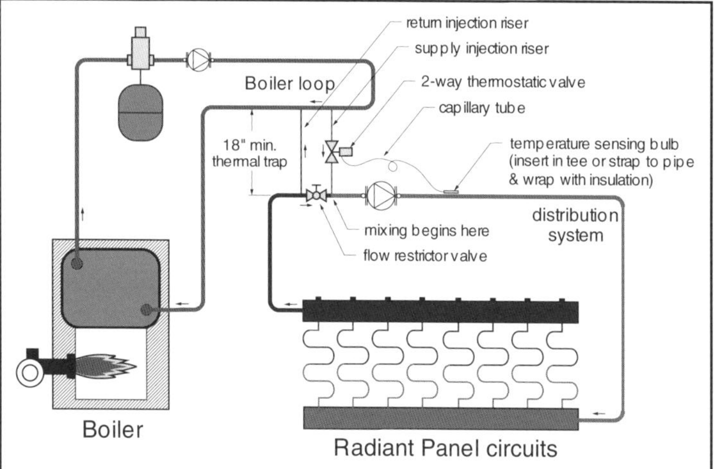

NOTE:

- The flow restrictor valve is partially closed (as necessary) to achieve the proper injection flow rate as the thermostatic valve opens. The more the flow restrictor valve is closed, the greater the injection flow rate, and the higher the supply temperature to the radiant panel circuits.
- The system shown above does not protect the boiler from flue gas condensation. A second temperature sensing valve is needed to ensure the boiler is protected. Some boilers provide internal return temperature protection.
- The thermal trap formed by the injection risers prevents hot water in the boiler loop from migrating into the distribution system through the return injection riser.

Figure 4-39 Injection Mixing with 2-Way Valve

4

This method of control provides only fixed (setpoint) control, not reset control.

The rate hot water flows through the 2-way thermostatic valve depends on the pressure drop between the tees connecting the injection risers to the distribution loop. This pressure drop, in effect, establishes the "driving force" to push hot water into the distribution loop as the 2-way opens. The greater this pressure drop is, the faster hot water will by injected into the distribution loop as the thermostatic valve opens. A "flow restrictor" valve (usually a globe valve or ball valve) installed between these tees allows the pressure drop to be adjusted as needed.

The flow restrictor valve is partially closed as necessary to achieve design supply water temperature to the radiant panels. Once set, the handle of the flow restrictor valve should be removed to prevent tampering.

Figure 4-40 describes how to properly set the flow restrictor valve based on temperature measurements. It's important to follow this procedure when starting and adjusting a high thermal mass radiant panel (such as a slab-on-grade floor system).

The function of the boiler loop is to provide a substantial bypass flow of hot water past the tees where the injection risers connect. The bypass flow mixes with the cool water from the return injection riser, boosting its temperature before entering the boiler. This boiler loop provides a degree of protection against flue gas condensation in a conventional boiler. It does

The purpose of the flow restrictor valve is to create sufficient pressure differential to force hot water through the 2-way thermostatic valve at the proper flow rate. To set this valve calculate the design temperature drop across the radiant panel circuits, then use the formula below to calculate the "start-up" temperature difference.

$$(T_s - T_R)_{start\text{-}up} = (T_s - T_R)_{design} \times \left(\frac{(T_i - T_R)_{start\text{-}up}}{(T_i - T_R)_{design}}\right)$$

where:
Ti = temperature of injection water available (F)
Ts = temperature of supply water to radiant panel circuits (F)
TR = temperature of return water from floor circuits (F)

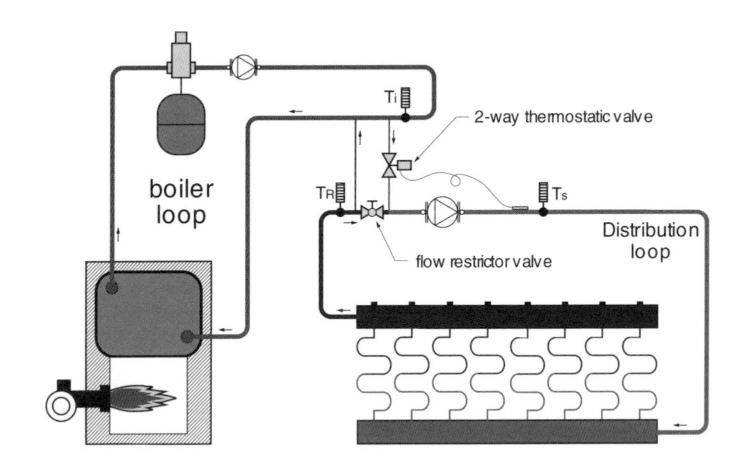

For example: Assume a floor heating system requires 110 F water at design conditions. During start-up, return water from the cool slab is at 60 F. At the same time boiler water is available for injection at 170 F. Assume the design temperature drop of the radiant panel manifold is 10 F. Thus the return temperature from the radiant panel cicuits at design conditions is 110-10=100 F. The temperature drop to set across the manifold at start-up is calculated using the formula:

$$(T_s - T_R)_{start\text{-}up} = (10) \times \left(\frac{(170-60)}{(170-100)}\right) = 15.7\text{ F}$$

With the 2-way thermostatic valve fully open, slowly begin closing the flow restrictor valve while monitoring the difference between temperatures Ts and TR. When this differential equals the calculated value the flow restrictor valve is properly set. The higher the injection water temperature and the lower the radiant panel return temperature, the *less* the flow restrictor valve needs to be closed. Note: Even after the flow restrictor valve is properly set, the supply water may not feel very warm. It will increase as the thermal mass of the radiant panel warms up.

Figure 4-40 Setting the Flow Restrictor Valve

not provide complete protection during transient operation (such as when warming a cold floor slab). To provide complete protection another thermostatic valve would have to be used to shunt boiler water directly back to the boiler return. In effect this valve "robs" hot water from the injection risers to give priority to protecting the boiler against condensation.

The injection riser piping should be installed with a vertical drop of at least 18" between the boiler loop and the distribution loop. This creates a thermal trap in the return injection riser, preventing hot water from migrating down into the distribution loop when no heat input is needed. The closed 2-way valve protects the supply injection riser from heat migration. It could be argued that when the boiler loop serves only the radiant panel system this thermal trap is not necessary (since the boiler and boiler loop circulator could be shut off). However, many hydronic systems are set up with multiple loads, each requiring hot water flow in the boiler loop. In these situations it's very likely that hot water will be flowing across the tees connecting the injection risers to the boiler loop at times when no heat input is required by the radiant panel. The thermal trap is critical in such cases.

Thermal traps with a vertical drop of at least 18" are needed to prevent unwanted heat migration into the distribution loop.

Injection Mixing using Variable Speed Pumps:

A small pump can also be used to force hot water from the boiler loop into the distribution loop. By varying the speed of the pump, the rate of hot water injection, and thus the temperature of the distribution loop can be regulated. This method of injection mixing relies on a controller specifically designed to adjust the speed of a small circulator. Controls are available for both conventional AC wet-rotor circulators and small DC-driven pumps. Such controls can be configured to provide either fixed temperature or reset control of the radiant panel.

There are two methods of piping a variable speed pump for injection mixing; one is called direct injection, the other reverse injection. Both have strengths and weaknesses depending on the intent of the system.

Direct (Variable-Speed) Injection Mixing:

A schematic showing variable speed (direct) injection mixing is given in Figure 4-41.

When a small wet-rotor circulator is used as the injection pump in a system with high temperature boiler water and low temperature return water, it's often capable of injecting heat at a rate much higher than the distribution system requires. This happens for two reasons: First, the required hot water injection flow in a low temperature radiant panel system is very small, even at design load conditions (see Formula 4-1. Secondly, because the injection risers are connected to both the boiler loop and distribution loop with closely spaced tees, the injection pump must only overcome the head loss of the short injection risers. The combination of these two factors limits the injection pump to a small percentage of its full speed, even under design conditions. This wastes a large portion of the controller's speed adjustment capability.

To correct for this, a globe valve is installed in the return injection riser to purposely increase the head loss against which the injection pump operates. This forces the pump to run over a wider portion of its speed range. Small DC-operated "micropumps" that operate on only a few watts of power do not need to have their injection flow restricted in this manner.

Direct injection mixing offers the greatest rate of heat transfer into the distribution system for a given injection flow rate and temperature. It is well suited for large residential and light commercial systems.

Two piping details that are crucial to the success of direct injection systems are the spacing of the primary/secondary tees and the formation of a thermal trap.

The spacing between the primary/secondary tees in both the boiler and distribution loops should be as small as possible (in no case greater than 4 pipe diameters). Copper tees are preferred because of their smooth inner surfaces. The tube stub connecting these tees should be carefully reamed and cleanly soldered to minimize any pressure drop between the tees. Such a pressure drop encourages hot water to migrate from the boiler loop into the distribution loop even when the injection pump is totally off. Since many radiant floor systems maintain continuous circulation through the floor circuits, this slow but persistent trickle of hot water will continually inject heat (albeit at a low rate) into the floor circuits even when the building doesn't need it and the injection pump is off. The

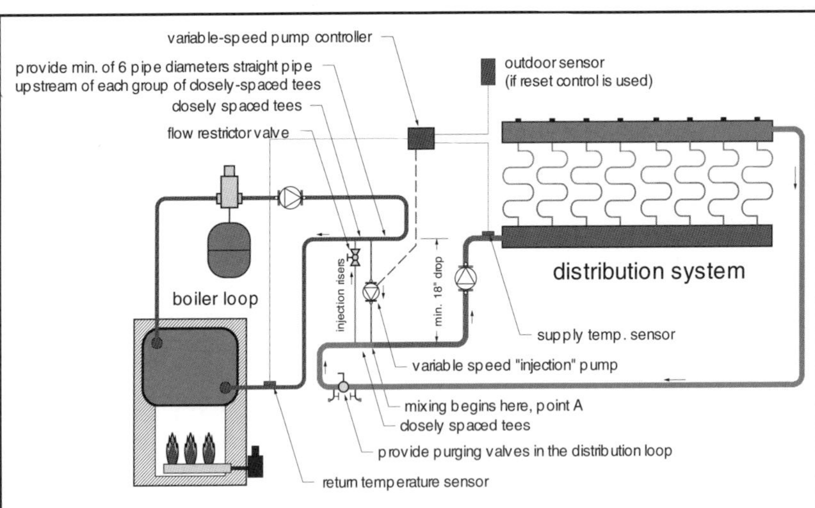

NOTE:

- The tees connecting the injection risers to the boiler loop should be spaced as close together as possible.
- The tees connecting the injection risers to the distribution loop should be spaced as close together as possible.
- It's crucial that the injection risers drop at least 18" vertically from the boiler loop to the distribution loop to form a thermal trap.
- Direct injection mixing yields a higher heat transfer rate (per g.p.m. of injection flow) than does reverse injection mixing.
- The distribution circuit can operate with either fixed temperature or reset water temperature.
- Mixing controls slows injection pump when necessary to protect boiler against excessively low return temperatures.

Figure 4-41 Direct Injection Mixing

result could be overheating during mild weather, especially in smaller systems.

The vertical thermal trap between the boiler loop and distribution loop discourages thermal migration when the injection pump is off. Both injection risers must have a minimum vertical drop of 18" to prevent hot water migration downward into the distribution loop.

It's also good practice to have at least 6 pipe diameters of straight pipe upstream of the first tee in any primary/secondary connection. This allows flow eddies created by fittings and valves to smoothen out before the flow enters the first tee.

The use of weighted (flow check) or spring-loaded check valves in the injection risers is NOT recommended because it leads to unstable injection pump operation under low-load conditions.

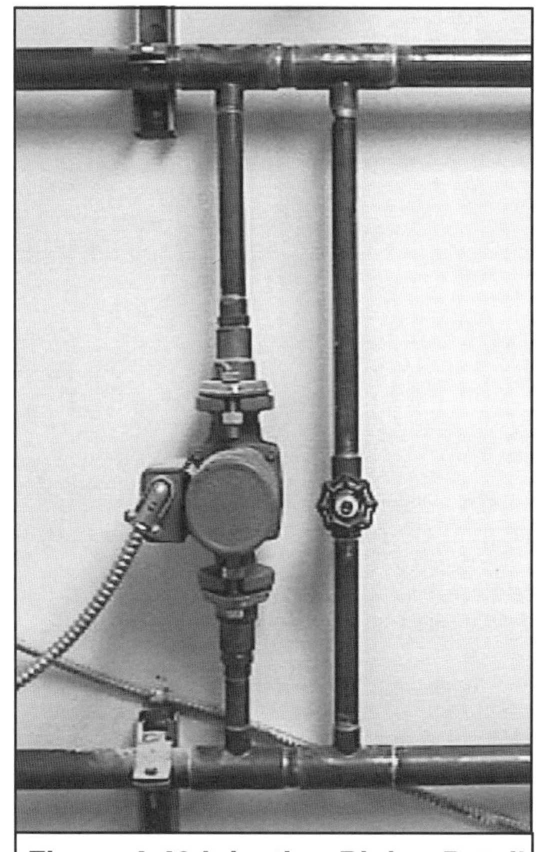

Figure 4-42 Injection Piping Detail

The piping details just discussed can be seen in the installation shown in Figure 4-42.

Although hot boiler water and cool return water begin to mix at the tee below the injection pump, mixing may not be complete for some distance downstream of this point. To ensure proper mixing it's preferable to install the supply temperature sensor downstream of the distribution circulator.

When piping a direct injection system it's important to incorporate a means of purging the distribution loop. Because the primary/secondary tees at the injection risers effectively decouple the boiler loop from the distribution loop it can be difficult and slow to purge the distribution loop by relying on purging flow in the boiler loop alone. To provide for effective purging, install a hose bib valve on either side of a full-port ball valve in the distribution loop as shown in Figure 4-41. During purging the ball valve is closed and water is forced into the downstream hose bib, around the loop, and eventually out the other hose bib. Another option is to install one hose bib valve and a full port ball valve a few pipe diameters upstream of the return injection riser. During purging the ball valve is closed and the makeup water system on the boiler loop is put into fast fill mode. This forces water around the distribution loop and eventually out the hose bib valve.

Reverse (Variable-Speed) Injection Mixing:

When variable speed injection mixing is used in a small, low-temperature ra-

diant panel system, reverse injection mixing offers some advantages over direct injection mixing. Figure 4-43 shows a system configured for reverse injection mixing.

There are two important differences in the piping used for reverse injection mixing compared to that of direct injection mixing. First, the return injection riser is connected so that water returned to the boiler loop is at the radiant panel supply temperature (rather than the return temperature as in direct injection mixing). This decreases the temperature differential between the incoming and outgoing injection risers, and, according to Formula 4-1, re-

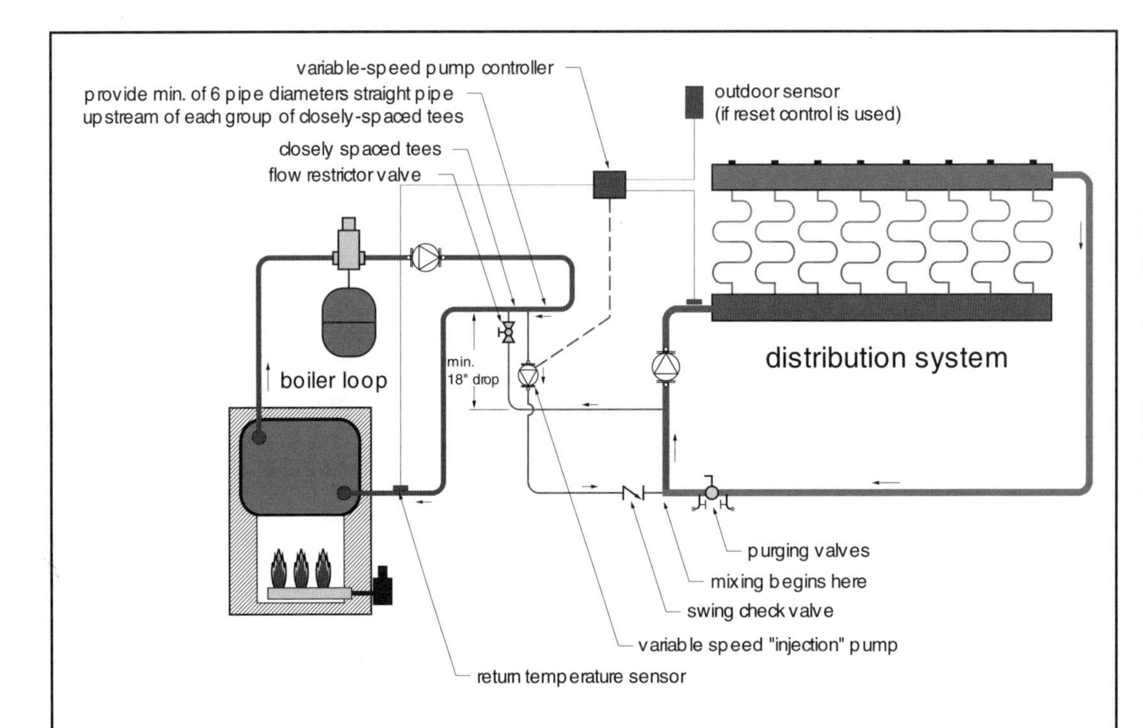

NOTE:

- The tees connecting the injection risers to the boiler loop should be spaced as close together as possible.
- Use only a *swing check* valve in the supply injection riser. Install piping in vicinity of the swing check exactly as shown above.
- Reverse injection mixing yields a lower heat transfer rate (per g.p.m. of injection flow) than does direct injection mixing.
- Distribution loop can operate with either fixed temperature or reset water temperature.
- Mixing controls slows injection pump when necessary to protect boiler against excessively low return temperatures.

Figure 4-43 Reverse Injection Mixing

quires a higher injection flow rate. Although this probably seems counterproductive, remember that the typical small wet-rotor circulator is already capable of providing more injection flow than required, and thus requires a considerable amount of its head to be throttled away by the flow-restricting valve. Reverse injection reduces the throttling requirement of the flow restricting valve.

The second piping difference is the swing check valve in the supply riser. This prevents any flow in the hot injection riser when the injection pump is off. Note: Only a swing check valve is acceptable here. Do not use a spring-loaded check valve or flow-check valve. It's also important to install the piping in the vicinity of the check valve exactly as shown in the schematic. This allows flow rounding the corner of the distribution loop at tee (A) to create a stagnation pressure to backset the check valve when the injection pump turns off. It's still necessary to create a vertical thermal trap in the return riser to prevent thermal migration.

Multi-Load Systems:

Many homes and commercial buildings require two or more radiant panel sub-systems that operate at different water temperatures. A house, for example, might use slab-on-grade floor heating on its ground floor, combined with a staple-up plate system on the second floor. Depending on design specifics these two sub-systems could require considerably different supply water temperatures.

Factors such as floor coverings and budget often lead to buildings that are not completely heated by hydronic radiant panels. Instead "hybrid" systems are designed that may use radiant panels in certain areas combined with other types of hydronic heat emitters or air handlers in other areas.

It's also a common requirement to supply a building's domestic water heating from the same hydronic heat source used for supplying the radiant panel(s).

All these requirements can usually be incorporated into a single hydronic system. The key is planning it correctly. Although this section cannot cover all design issues in detail, it does describe the concept of primary/secondary piping as a "framework" upon which to build even sophisticated multi-load systems.

Primary/Secondary Piping:

An important factor in supplying several heating loads from a single system is eliminating "flow interference" between the loads. For example, if several circulators, some large and some small, are piped to a single header system connected to a boiler, operation of a large circulator could seriously interfere with simultaneous operation of a smaller circulator. This in turn can lead to erratic heat delivery to the radiant panels (or other hydronic heating loads).

Primary/secondary piping provides a means of interfacing several hydronic distribution circuits to a common heat source without creating such flow interference problems. The basic idea is shown in Figure 4-44.

In a true primary/secondary piping system the primary circulator must operate whenever any of the secondary loads require heat. The closely spaced tees that connect each secondary circuit to the primary circuit prevent flow interference, regardless of which circuit has the greater flow rate. It's crucial that these tees are close together. The spacing between their side ports should not exceed 4 times the diameter of the primary piping. Copper tubing and tees are preferred over threaded pipe at primary/secondary connections because the later creates greater turbulence between the tees.

It's possible to add multiple secondary circuits onto a common primary loop as shown in Figure 4-45. Secondary circuits that require higher water temperature should be located closer to the boiler outlet than those capable of operating at lower water temperatures. For example, in Figure 4-45 the secondary circuit supplying the panel radiators is located upstream of the one supplying the floor heating circuits. The designer should remember that water temperature in the primary circuit drops whenever it passes the primary/secondary tees of an operating secondary circuit. If several secondary circuits are operating simultaneously there could be a substantial drop in water temperature by the time the primary flow makes it to the last secondary circuit. This is not necessarily a problem, but it does require the designer to size the downstream loads accordingly.

Another way of connecting multiple secondary loads is shown in Figure 4-46. Here the primary circuit is broken up into parallel branches, each of which serves a secondary load. This design provides the same temperature water to each secondary load. Although more piping is involved, it may allow smaller heat emitters to be used

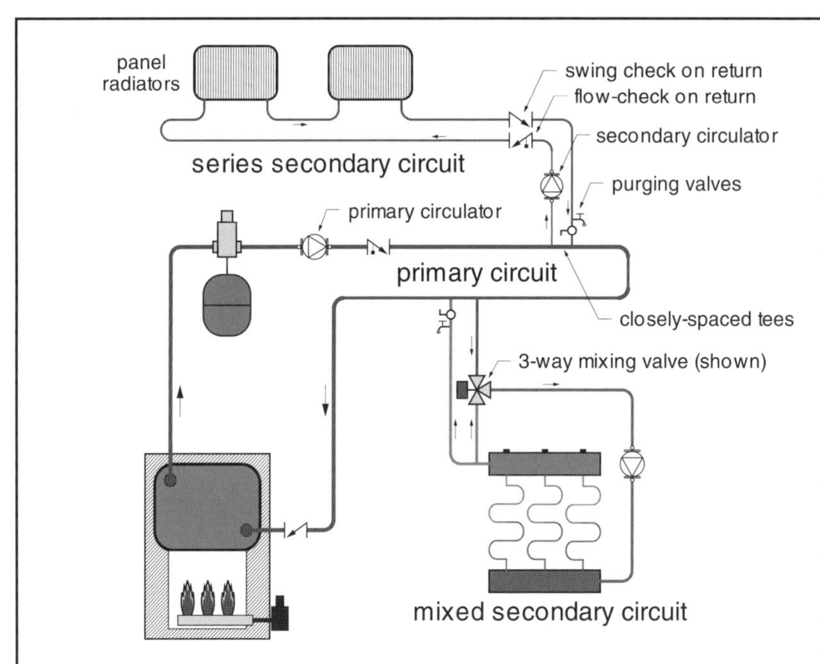

NOTE:

- Flow in primary circuit will not "induce" flow in secondary circuit.
- The center-to-center spacing of the primary / secondary tees should not exceed 4 time the larger pipe diameter.
- Allow at least 6 pipe diameters of straight pipe upstream of the first tee, and at least 4 diameters of straight pipe downstream of the second tee, at all primary / secondary connections.
- All secondary circuits (other than those for injection mixing with variable speed pumps) should have a flow-check valve on the supply pipe to prevent heat migration.
- All secondary circuits (other than those for injection mixing with variable speed pumps) should have a flow-check or swing check valve on the return pipe to prevent heat migr
- Provide purging valves in each secondary circuit.

Figure 4-44 Primary/Secondary Piping

4

in part of the system.

Although it's possible to supply an indirectly fired domestic water heater as one of several secondary loads on a common primary circuit, there may be reasons not to do so. For example, if the primary circuit is long, or travels outside the mechanical room, it will give off heat to the building anytime domestic water needs heating. The heat loss of 100 feet of 1.5" uninsulated copper tubing carrying 180 °F water through a space at 75 °F is about 8500

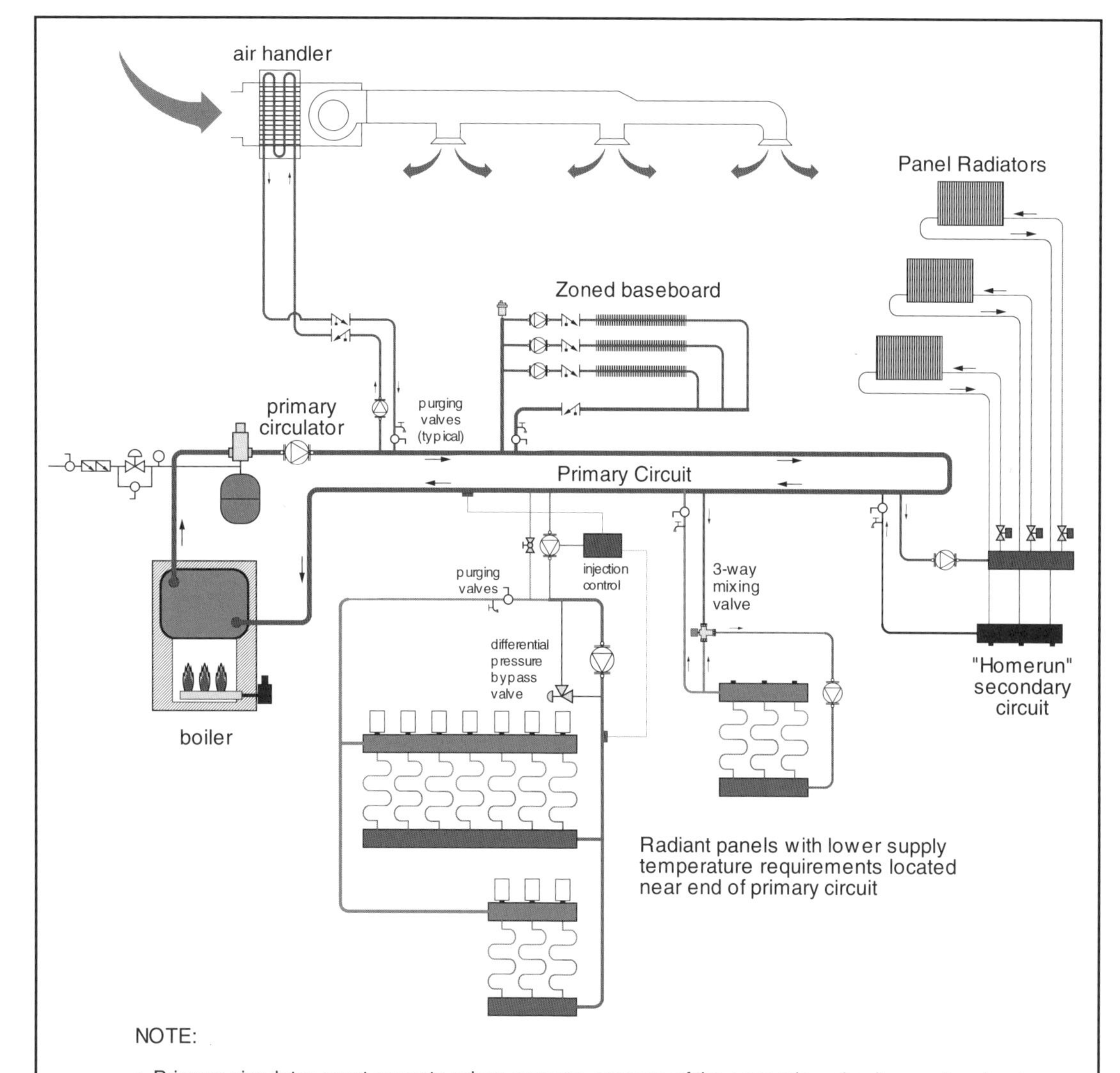

NOTE:

- Primary circulator must operate when any one or more of the secondary circuits requires heat.
- Secondary circuits should be arranged around the primary circuit in decending order of required supply temperature.
- Each secondary circuit is provided with valves for purging.
- Boiler return temperature protection is provided only by the injection mixing control. Other secondary circuits could cause lower boiler return temperature during cool start-up conditions, or simulataneous operation of all secondary loads.

Figure 4-45 Multi-Load System with Common Primary Circuit

Btuh. When this occurs, as it will, during hot summer days, it adds almost 3/4 ton to the home's cooling load. Short, well-insulated primary circuits reduce this extraneous heat loss. Another way to avoid this situation is to pipe the DHW tanks as a parallel rather than secondary load as shown in Figure 4-47. Be sure to install a flow-check valve in each parallel circuit to prevent reverse flow. The DHW tank can be controlled as either a priority or non-priority load depending on how the controls are configured.

Another important aspect of primary/secondary piping is protecting the secondary circuits against thermal migration. The basic idea is simple: Hot water wants to rise in the system. It will do so if given the opportunity (even through a single pipe). Although the gravity-induced flow is relatively weak compared to flow generated by circulators, it none the less can create noticeable heat output in any kind of heat emitter when it's least wanted (like on a hot July afternoon).

There are several ways to protect secondary circuits against thermal migration:

- Install a flow-check valve on both the supply and return risers when they connect into the top of the primary circuit.
- Provide a flow-check valve on the supply riser and create an underslung thermal trap on the return riser.
- Create a deep thermal trap in both supply and return risers.

When planning the layout of a primary/secondary piping system, it's preferable to keep the primary circuit piping up high in the mechanical room and drop down from it to begin a secondary circuit, (even if the secondary circuit has to turn and go back up into the building). This inherently builds thermal traps into the secondary circuits.

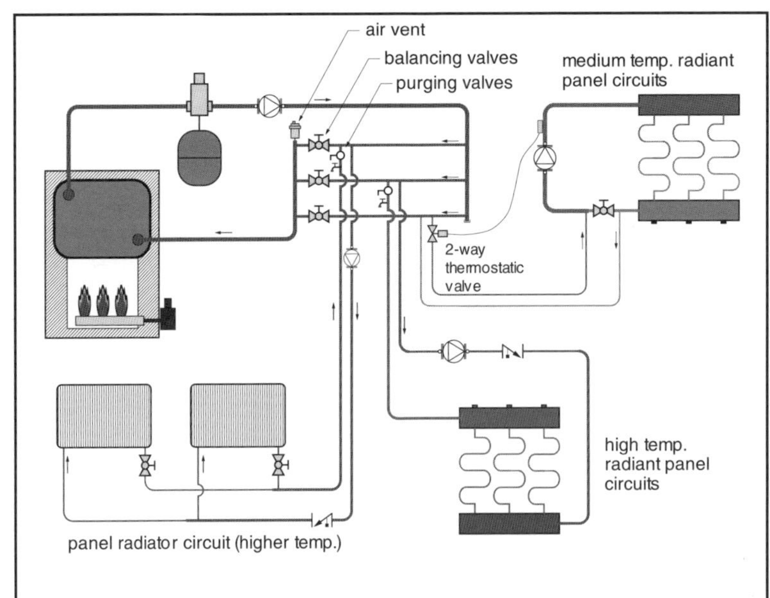

NOTE:

- Each secondary circuit receives the same water temperature from the (split) primary circuit.
- Seconday circuits can be high temperature (direct hot water feed), or low temperature (using a mixing assembly).
- Flow balancing valves should be installed in parallel segments of primary circuit if secondary circuit loads are significantly different.

Figure 4-46 Parallel Branch Primary Circuit

Design Summary for Primary/ Secondary Systems:

1. Keep primary/secondary tees no more than 4 pipe diameters apart.
2. Use smooth tees and smoothly reamed tube stub between tees.
3. Allow at least 6 pipe diameter of straight pipe ahead of each up-stream tee at all primary/ secondary connections.
4. Protect secondary circuits against thermal migration.
5. When multiple secondary loads are arranged in sequence, size down-stream loads for reduced water temperature.
6. Provide parallel branches in the primary circuit if all secondary circuits require the same supply temperature.
7. Provide a means of purging each secondary circuit at start-up.

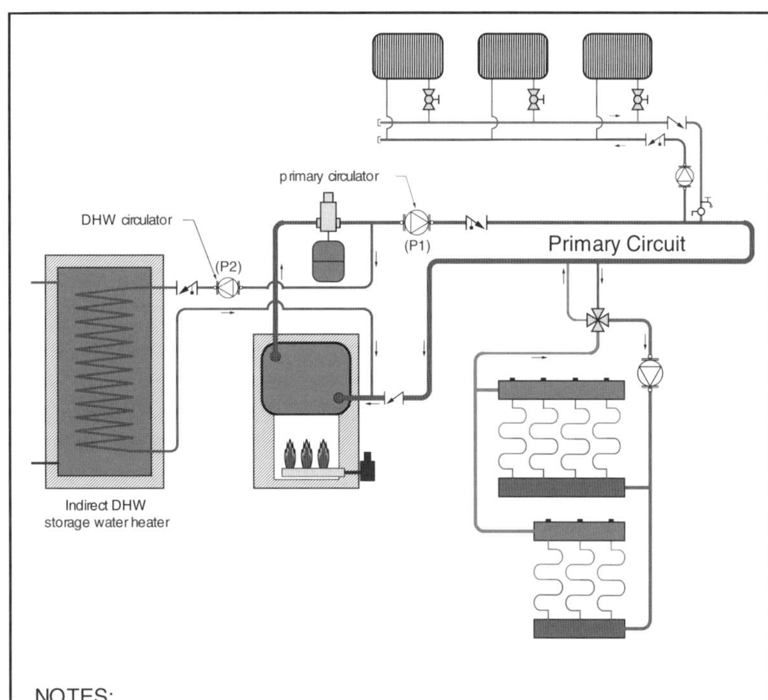

NOTES:

• Piping the DHW tank as a separate circuit eliminates the need to operate the primary loop for domestic water heating

• Flow-checks are required in *both* the primary circuit and DHW circuit to prevent reverse flow through inactive circuits. They also prevents gravity flow in either circuit.

• DHW can be operated as a "priority" load by wiring controls so that the primary circulator (P1) is off whenever DHW circulator (P2) is on.

Figure 4-47 Parallel Piping of DHW Tank

4•6 Hydronic Heat Sources

A wide variety of heat sources can be used with hydronic radiant panel heating systems. These include gas- and oil-fired boilers, hydronic heat pumps, and domestic water heaters to name a few.

In fact, almost any source of warm water is a potential heat source for a hydronic radiant panel system. The basic characteristics of several types of hydronic heat sources are described in this section. More detailed information pertaining to their selection and installation is best obtained from manufacturer's literature and manuals.

Gas- and Oil-fired Boilers:

Although there are many terms used to describe modern gas- and oil-fired boilers, ("sealed combustion", "power-vented", "high mass", "low mass", and more), perhaps the most important classification in the context of radiant panel heating is whether the boiler is "conventional" versus "condensing".

The word conventional refers to a gas- or oil-fired boiler that's intended to operate without condensation of flue gases inside the boiler. Such boilers are commonly used for hydronic heating systems using fin-tube baseboard convectors, panel radiators, or fan-coils. They typically have heat exchangers constructed of cast-iron, steel, or finned copper tubing. With proper piping and controls conventional boilers can supply heat to almost any kind of radiant panel system.

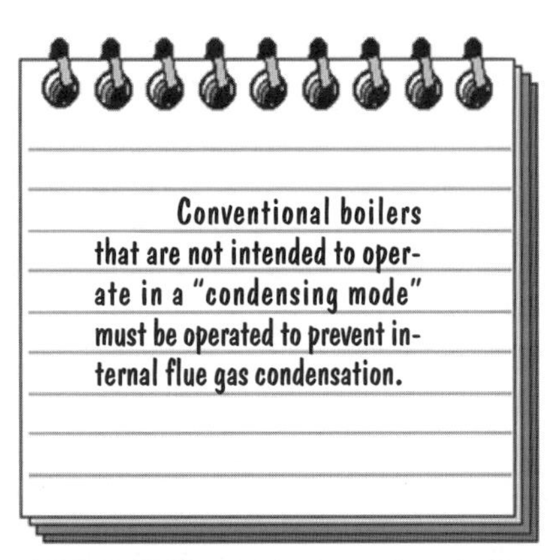

Condensing boilers, by contrast, are specially designed to cause flue gases to condense inside the boiler. In doing so they recover the latent heat contained in the exhaust gases, and can attain higher fuel efficiencies than conventional boilers.

Flue Gas Condensation:

All boilers must deal with flue gas condensation in one way or another. It's important to understand the principles involved, as well as the potential problems that can occur if flue gas condensation is not properly addressed during system design.

As the temperature of the exhaust gases produced inside a boiler decreases, a point is reached where water vapor, (a by-product of combustion), as well as other chemical vapors in the exhaust stream, condense into liquids.

Depending upon the design and operating conditions of the boiler, condensation can occur inside the boiler, inside its vent piping, inside the chimney, or not until the exhaust gases exit the chimney and mix with cold outside air. The condensate that forms is highly acidic and can quickly corrode materials such as steel, cast-iron, and even copper. It can also cause deterioration of masonry chimneys.

A similar situation takes place in the exhaust system of a car on a cold winter morning. When the car is first started, its exhaust system is cold. Under such conditions, some of the water vapor produced by combustion in the engine condenses into a liquid on its way through the exhaust system. Shortly after the car is started the condensate can be seen dripping out the tailpipe. After the exhaust system has warmed up the water vapor usually makes it all the way down the tail pipe without condensing, and thus the tailpipe stops dripping. Anyone who lives in a cold climate,

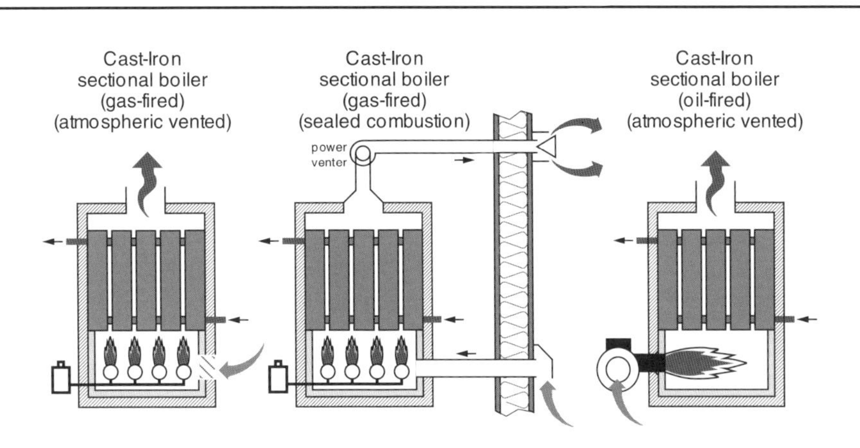

- Not intended for condensing mode operation
- Minimum return water temp. requirements
- Higher thermal mass
- Draws combustion air from room. Requires chimney or suitable power venting system.

- Not intended for condensing mode operation
- Minimum return water temp. requirements
- Higher thermal mass
- Draws combustion air from outside. Requires approved air supply and venting piping

- Not intended for condensing mode operation
- Minimum return water temp. requirements
- Higher thermal mass
- Requires chimney or suitable power venting system

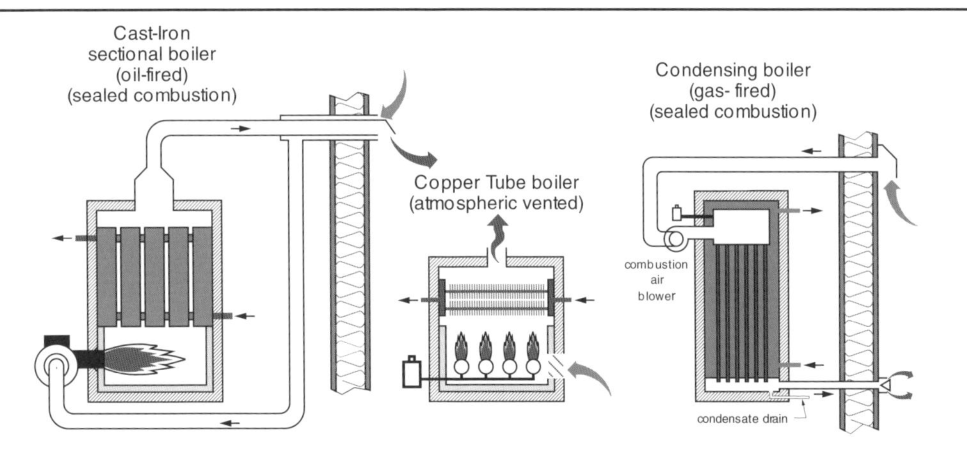

- Not intended for condensing mode operation
- Minimum return water temp. requirements
- Higher thermal mass
- Draws combustion air from outside. Requires approved air supply and venting piping

- Not intended for condensing mode operation
- Minimum return water temperature requirements
- Low thermal mass
- Requires chimney or suitable power venting system

- Designed to operate in condensing mode for highest efficiency.
- The lower the return water temp. the higher the efficiency.
- Can be either high or low mass depending on design.
- Requires drain tube for condensate

Figure 4-48 Gas- and Oil-Fired Boiler Options

4

and makes short wintertime trips in their car likely knows the effect on their exhaust system. Steel tailpipes and mufflers experience rapid corrosion under such condition, often failing in as little as one year.

Visualize the same situation inside a boiler and its flue piping. If the boiler is filled with relatively low temperature water, (such as that returning from a typical low temperature floor heating system), the water vapor produced during combustion will condense as it makes contact with the boiler's heat exchanger. Severe internal corrosion and premature boiler failure can result. The situation is even more ominous considering the fact that when flue gas condensation forms inside the boiler, it's also likely to be present in the flue pipe. The typical light gauge galvanized steel vent piping used with conventional boilers is no match for the acidic condensate that can eat holes through it after only a few weeks of operation. This in turn can cause leakage of carbon monoxide inside the building - a potentially life-threatening situation.

The key to avoiding flue gas condensation is to ensure that return water temperature stays above the dewpoint temperature of the water vapor in the exhaust gases. For a typical gas-fired boiler this temperature is around 130 °F. For oil-fired boilers the dewpoint temperature of water vapor is lower, (due to lower hydrogen content in the fuel). However, some fuel oils contain higher amounts of sulphur that can lead to formation (and condensation) of sulfuric acid at return water temperatures below 150 °F.

The piping designs presented in Section 4-5 show several ways of boosting return water temperature to prevent sustained flue gas condensation in the boiler.

Condensing Boilers:

Not all characteristics of flue gas condensation are undesirable. Anytime water vapor condenses back to a liquid there is a considerable amount of heat released in the process. For each therm (100,000 Btu) of natural gas burned, approximately 9.6 pounds of water vapor is produced. If all of this vapor condenses inside the boiler, it releases an additional 9300 Btu of heat from the exhaust stream. That's over 9% more heat from the same amount of fuel. Heat that is carried out with the exhaust gases in non-condensing boilers.

Two requirements must be met to successfully reap the benefits of high efficiency/condensing mode operation. First, the internal components of the boiler must be designed and built to withstand the acidic nature of the condensate without corrosion. Second, the boiler must be part of a system that consistently returns low temperature water to the boiler, thus ensuring sustained flue gas condensation.

The first objective has been met by a number of boiler manufacturers. There are several fully condensing boilers presently available on the US market. Their heat exchangers are typically constructed of high-grade stainless steel that can withstand the acidic condensate without corroding.

The second requirement depends on the system the condensing boiler is installed in. Systems that operate at low return water temperatures such a low-temperature radiant panel heating, or snow melting, are good matches for condensing boilers. In such applications a condensing boiler can have sustained operating efficiencies in the mid 90% range. Other types of radiant panel systems, such as suspended tube floor heating, or some types of radiant wall pan-

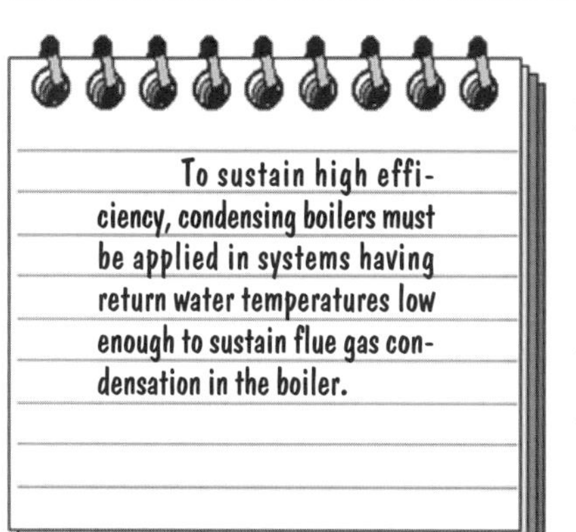

els operate at higher water temperatures, and do not return water temperatures to the boiler low enough to sustain condensing mode operation. Under such conditions, a condensing boiler will operate at efficiencies comparable to those of conventional boilers.

During a typical winter's day, several gallons of condensate can be produced. It must be routed to a suitable drain. Floor drains made of PVC are usually acceptable. Most municipal plumbing codes allow disposal of the condensate into sanitary sewers, provided a suitable trap is placed between the boiler and drain line.

However, be advised that the disposal of acidic condensate can cause deterioration of old clay drainage tile over time. This has occurred in Europe, where codes now require condensate to be chemically neutralized before being released into a public sewer system.

Never allow the condensate to simply run out onto a concrete floor. In time it will deteriorate the concrete as well as corrode other metals it comes in contact with. If a floor drain is not available, the condensate must be collected in a sump and pumped out using a float-switch operated condensate pump.

Condensing boilers require different installation materials and methods from those used with conventional boilers. Supply air for combustion is often supplied directly to the boiler through a PVC pipe routed out through the wall of the building. Similarly, the relatively cool exhaust gases are exhausted through a separate CPVC pipe. The low temperature exhaust gases (often under 150 °F) must be forced through the exhaust system by a special blower built into the boiler. The low temperature exhaust gas will not create the draft necessary for chimney venting. Because their operating characteristic are considerably different from conventional boilers it's crucial to follow the condensing boiler manufacturer's installation specifications.

Hydronic Heat Pumps:

The vast majority of currently installed residential heat pumps use a standard refrigeration cycle to gather heat from outside air, upgrade the temperature of this heat, and then release it into a building via a forced-air delivery system. In recent years however there has been a considerable increase in the use of heat pumps that gather heat from the soil rather than outside air. The key advantage being that soil temperature a few feet below the surface is often significantly higher than that of the outside air, especially in cold northern climates. The higher the temperature of the "source" of heat, the greater the efficiency of any heat pump. In some cases these so-called "geothermal" heat pumps can have efficiencies 50% to 100% higher than conventional air-to-air heat pumps.

A water-to-water (hydronic) heat pump can extract heat from the soil and deliver the heat at a higher temperature to a stream of water flowing through its condenser. Such a unit is well suited as a heat source for low temperature radiant panel applications. This section describes the general characteristic of hydronic heat pumps. Again it's best to consult the manufacturer for specific performance information and installation requirements.

Open Loop Heat Pump Systems:

One method of gathering heat from the earth is to circulate water from a well, large pond, or lake, through a water-to-refrigerant heat exchanger within the heat pump. Such water is often between 35 and 60 °F depending on its source, geographic location, and time of year. As the water flows through the heat pump its temperature drops 5 to 15 °F as heat is extracted. The water then returns to its source through a separate pipe where it warms up to its original temperature. Such an arrangement is called an "open-loop" geothermal heat pump system. Two factors are critically important to the success of open-loop systems.

First, there must be a source of water near the building that can continually supply the heat pump hour after hour, day after day. A typical residential heat pump system requires 6 to 12 gallons per minute of sustained water flow whenever it's operating. On a cold winter day this results in several thousand gallons of water passing through the system in a 24-hour period. If the source of water, (a marginal well for example), cannot maintain this water demand, the heat pump will eventually shut itself off due to low refrigerant suction pressure. This can leave the building without heat, and perhaps even worse, out of water (if the heat pump and domestic water come from the same well). If you're considering an open-loop geothermal heat pump as the heat source for a radiant panel system, be absolutely certain the water source can supply the required quantity of source water. Have a certified well driller verify the sustained recovery rate of any drilled well being considered.

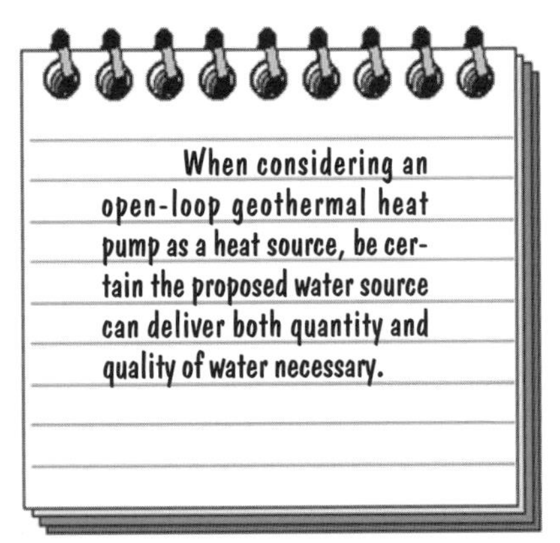

Secondly, the water supplied to the heat pump also must be of sufficient quality. Water containing calcium carbonate, iron bacteria, hydrogen sulfide, or other mineral or bacterial materials can create major problems for the heat pump. Such material will eventually clog the water-to-refrigerant heat exchange in the heat pump, again causing a water starvation condition and automatic shutdown. Cleaning the heat exchanger is messy, expensive, and usually only results in a temporary solution to the problem. Any water source being considered for use with an open-loop heat pump system must be thoroughly tested for mineral and bacterial content, by a quality testing laboratory. Be sure the levels of any contaminants are within the limits specified by the heat pump manufacturer.

Closed-Loop Heat Pump Systems:

Another way to extract heat from the earth is to circulate water, or a mixture of water and antifreeze, through a closed piping loop buried several feet below grade. Such a loop is constructed of high-density polyethylene or polybutylene tubing. It begins and ends at the heat pump. In combination with the heat pump's water-to-refrigerant heat exchanger it forms a closed-loop system. Once filled with water, and purged of air, nothing enters or leaves the loop.

Heat is extracted by chilling the water in the loop to a temperature lower than that of the surrounding soil. Heat migrates from the soil, through the wall of the tubing, and into the cooler circulating fluid. The fluid carries the heat back to the heat

pump where it's again extracted by a standard refrigeration process. The chilled water then flows back out into the buried loop to repeat the process.

Several US manufacturers offer water-to-water heat pumps suitable for use in radiant panel heating systems. Several single-phase models cover a capacity range from about 20,000 Btuh to over 60,000 Btuh. Larger capacity 3-phase models are also available for commercial applications.

Performance of Hydronic Heat Pumps:

It's important to understand that heat output from a hydronic heat pump is very dependent on the temperature of its "source" water as well as its "load" water. The load water is the return water temperature from the hydronic distribution system. Two principles apply to all hydronic heat pumps:

- The warmer the source water, the higher the heat output and efficiency of the heat pump.
- The lower the temperature of the hydronic distribution system, the higher the heat output and efficiency of the heat pump.

The variations in capacity and efficiency can be very significant. A water-to-water heat pump with a given heating capacity under favorable water temperatures, can have less than half this capacity under more demanding conditions. The heating capacity of water-to-water heat pumps is often listed in tables in manufacturers' literature over a wide range of entering water conditions. Selection of a hydronic heat pump must be based on the source and load temperatures that will occur in a particular project. Such conditions can vary considerably from the "standard" ARI rating conditions listed in manufacturers' literature. The graph in Figure 4-49 can help estimate performance variation over a range of operating conditions.

The fact that many hydronic radiant panel systems operate at relatively low water temperatures makes them a good

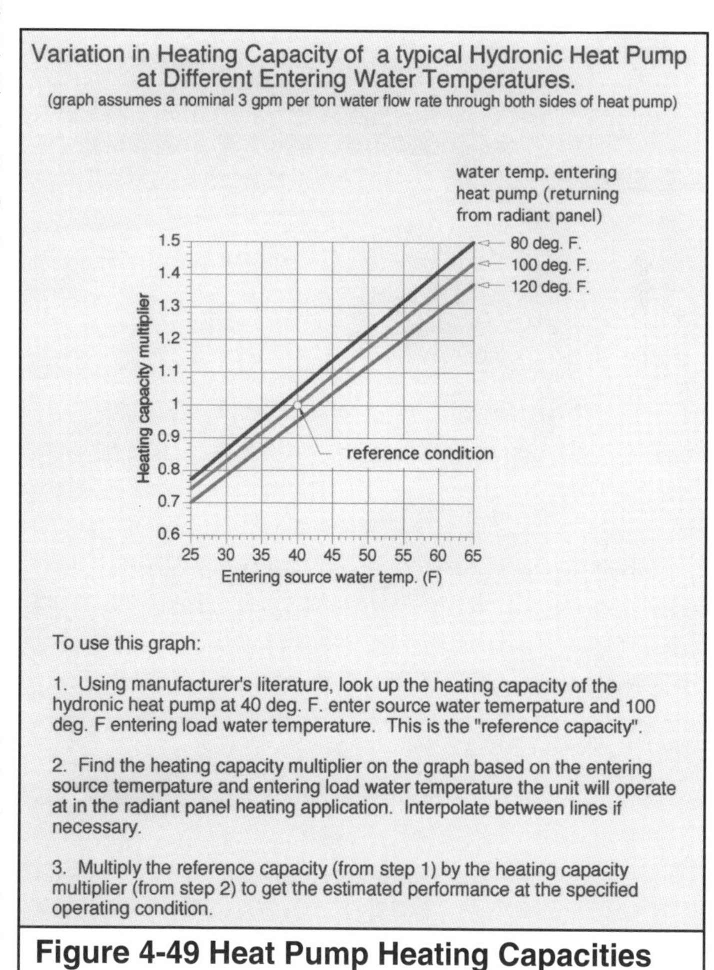

Figure 4-49 Heat Pump Heating Capacities

match for hydronic heat pumps. The lower the temperature at which the floor heating system can operate, the higher the heating capacity and efficiency of the hydronic heat pump supplying it will be.

Hydronic heat pumps always produce more heat output than the electrical energy required to operate them. The ratio of a heat pump's heating capacity, to the electrical energy required to run it is called its Coefficient of Performance, or COP for short. It can be calculated as follows:

$$COP = \frac{H_c}{(kw) \times 3413}$$

Formula 4-2

where

H_c = heating capacity in Btuh

kw = the electrical input to the heat pump (in kilowatts)

The higher the COP of a heat pump, the more efficient its operation. In other words, the more heat it produces using a given amount of electrical energy. Water-to-water heat pumps, connected to geothermal sources, and low temperature radiant panel heat delivery systems typically have COPs in the range of 2.5 to 3.0.

Buffer Tanks:

Heat pumps require constant water flow through both the evaporator and condenser heat exchangers while operating. If either flow stream is reduced beyond a certain point the heat pump automatically shuts itself off due to refrigerant pressure limits. This type of flow starvation can create several problems when a hydronic distribution system having several independently controlled zones is piped in series with the heat pump itself.

For example, imagine a situation where a given zone requiring 5000 Btuh calls for heat and in so doing turns on a 50,000 Btuh heat pump. All other zone valves remain closed. Two problems will quickly develop.

First the much higher heat input rate of the heat pump relative to the demand of the zone will force water temperature to climb rapidly. If a high limit control is used in the system, it will turn the heat pump off after only a few minutes operation. Short-cycling will persist, eventually shortening compressor life. If no high limit control is present, the heat pump will drive water temperature up to the point of turning itself off based on refrigerant pressure limits. Each time this happens the heat pump has to be manually reset before starting up. Obviously this is not a practical scenario.

Another problem develops because the flow rate required by the heat pump when operating can be substantially higher than the flow through the active heating zone. The long tubing circuits in most radiant panel systems will tend to "bottleneck" flow through the heat pump. The result will again be an automatic shutdown based on high refrigerant pressure.

The solution to this mismatch between heat output rate and zone heat demand is the inclusion of a "buffer tank" between the heat pump and the distribution circuits. The buffer tank is simply an insulated water storage tank, which provides the thermal mass necessary to "soak up" the extra heat output generated by the heat pump while simultaneously delivering heat at a lesser rate to an active zone. In effect the buffer tank "uncouples" heat production from heat distribution. It allows the heat pump to have reasonably long operating cycles, while at the same time allowing several independent zone circuits to distribute heat as required by the building. Figure 4-50 shows the placement of a buffer tank in the system.

Buffer tank sizing depends on several factors including the minimum acceptable on-time cycle of the heat pump, heat pump output, and the temperature differential of the tank's control. The following equation can be used to estimate the required volume of the tank:

$$V_{buffer} = \frac{(Q_{heatpump} - Q_{zone}) \times t}{500 \times \Delta T}$$

Formula 4-3

where:

V_{buffer} = required volume of buffer tank (gallons)

$Q_{heatpump}$ = heating capacity of heat pump

Q_{zone} = heat demand of smallest zone

ΔT = temperature differential of tank between heat pump ON and OFF cycle

t = minimum acceptable ON time for a heat pump cycle (minutes)

Example: What volume buffer tank is needed in a system with a 50,000 Btuh heat pump that should run at least 10 minutes when on? The smallest zone served by the tank is 5000 Btuh. The heat pump turns on when the tank reaches 100 °F, and off at 120 °F.

Solution: The ΔT of the tank is 120-100 = 20 °F Putting the numbers into the formula yields:

$$V_{buffer} = \frac{(50{,}000 - 5{,}000) \times 10}{500 \times 20} = 45$$

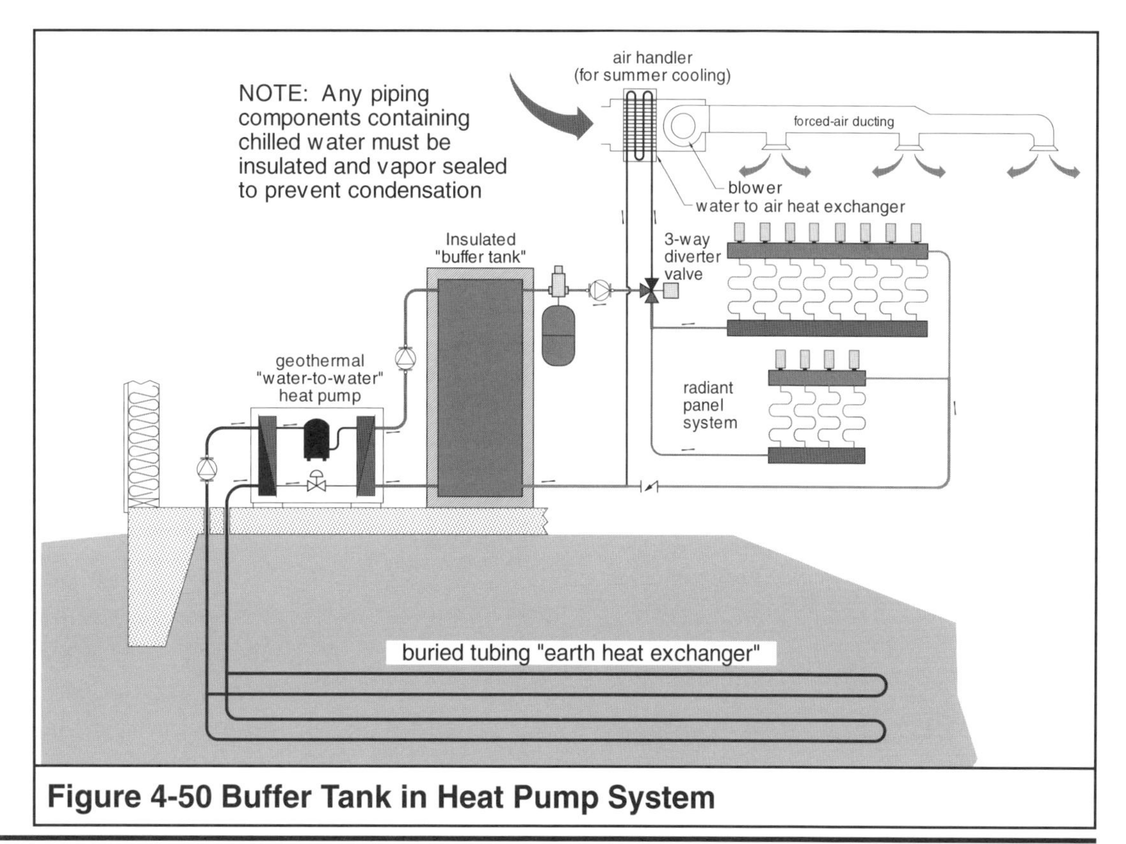

Figure 4-50 Buffer Tank in Heat Pump System

4

Cooling Option:

A unique feature of most hydronic heat pumps is the ability to produce chilled water for warm weather cooling. A reversing valve inside the heat pump switches the function of the refrigerant-to-water heat exchangers. Water circulating through the distribution system is now chilled rather than heated. The extracted heat is transferred to the water circulating through the buried earth loop where it's rejected to the soil.

The same buffer tank used in the heating mode can be used for storing chilled water provided it is insulated and the insulation is vapor sealed on the exterior side to prevent condensation. The chilled water can be circulated through one or more air-handlers or fan-coil units to both cool and dehumidify the building's air. These air-handlers or fan-coils must have a condensate drip pan and suitable drains to dispose of condensate. All piping carrying chilled water must also be insulated and vapor sealed to prevent condensation.

Another possible option is to route tempered chilled water through the same radiant panel circuits used for winter heating. It is however critically important that the panels remain a few degrees above the room's dewpoint temperature at all times to avoid the formation of condensation. Careful engineering is required on all radiant panel cooling projects.

The following is a summary of the advantages and disadvantages of hydronic heat pumps as heat sources for radiant panel heating applications.

Advantages:

- All electric. Gas service to the building may not be needed.
- No combustion. (No chimney or combustion air required. No exhaust gases)
- Ability to provide chilled water for cooling, if a suitable air-handler is used.
- No mixing valve required (less expensive controls)
- Installation cost rebates available from some electric utilities

Disadvantages:

- Must have suitable water source (for geothermal applications)
- Tend to make more noise than a modern boiler of equivalent capacity
- Tend to be more expensive than a gas- or oil-fired boiler of equivalent capacity
- Increase metered demand charges when used in a commercial building
- May require more specialized installation skills for earth loop
- Require refrigerant recovery if refrigeration system needs servicing

Domestic Water Heating Tanks:

Standard domestic water heating tanks can serve as a heat source for a radiant panel heating system under some circumstances. There are, however, several differences in comparison to gas- and oil-fired boilers that must be recognized and dealt with.

To begin with, most residential water heaters have heating capacities smaller than boilers. This often limits their use to smaller radiant systems. Multiple water heaters could also be used to build up capacity for larger loads.

Electric water heaters usually have heating elements of 3.8, 4.5, or 6.0 kilowatts. Converting these numbers into Btuh yields heat outputs of 12,970, 15,360, and 20,500 Btuh respectively. These relatively low heat outputs limit the viability of electric water heaters to small additions, or perhaps one or two individual rooms. Keep in mind that even though most electric water heaters have two heating elements, only one element operates at a time.

Some specialized water heaters have higher outputs and condensing capabilities for higher fuel efficiency. These water heaters are capable of heating small homes.

Residential gas water heaters typically have heat output limits of 30,000 Btuh. Such an output could likely handle an apartment or modest addition to a home depending on location and construction.

Oil-fired water heaters have the greatest heating capacity, (due to limitations on the minimum size of the burner nozzle). Output capacities of 50,000 - 60,000 Btuh are typical.

If the water heater has to provide both domestic hot water and space heating, the limited capacity becomes even more of an issue, especially considering the fact that high demand for hot water often occurs during wake-up time when space heating loads are also high.

It's possible to build a" priority" control system for such applications that would temporarily interrupt space heating when domestic water demand is high. This approach could work well in conjunction with a high-mass (slab-type) floor heating system, but could also prove unacceptable in low mass radiant panel systems that store very little thermal energy for "coasting" through periods of heat interruption.

Possible Piping Configurations:

Several possible piping configurations exist for use of domestic water heater tanks in hydronic radiant panel applications. These include:

- The tank is "dedicated" to space heating only
- The tank provides both DHW and space heating using the same water for both
- The tank provides both DHW and space heating but separates potable water from the space heating circuits.

Dedicated Water Heater Tank:

In this type of system a water heater is essentially substituted for a boiler. It must have all the same trim and safety controls that a boiler would have in the same installation. This includes a 30 psi pressure relief valve to protect of components in the system having lower pressure limits than the tank. It also includes a makeup water system, expansion tank, air separator, and purging valves. Some state or local codes may also require a low-water cutoff and manual-reset high limit control.

Because the tank serves only a space heating load it should be labeled as such, especially considering that someone might someday add antifreeze or corrosion inhibiting chemicals to the system. Ideally the tank should be located within heated space so that jacket heat loss contributes toward the building's heating load.

Combined DHW & Space Heating Using Potable Water :

Perhaps the most controversial issue associated with using a domestic water heater as a hydronic heat source focuses on circulating potable water through space heating circuits. In addition to the limited

4

capacity issue discussed earlier, several other potential problems could develop.

If the local water is hard, contains sediment, or is otherwise considered "aggressive", it could cause buildup or corrosion on components like circulators, valves, as well as the piping itself. This not only reduces heat transfer efficiency, it may also lead to premature component failures.

Because fresh water is always moving through the system, oxygen is always being resupplied to "feed" corrosion reactions with any ferrous metals in the system. To avoid such corrosion, all system components that come in contact with potable water must be non-corrodible. This includes major components like circulators, air separators, and the expansion tank, as well as all piping, fittings, and valves. In some cases this will add considerable cost to the system.

Concern has been raised that domestic water sitting stagnant in portions of the space heating piping during non-heating months could allow growth of bacteria. One proposed remedy is to install a timer that periodically circulates water through the space heating circuits to avoid stagnation. Some plumbing codes may specifically mandate this.

Lead-based solders, commonly used for copper tubing connections in hydronic space heating systems cannot be used on any piping containing potable water. Many codes also require type L copper tubing for domestic water piping versus the less expensive type M tubing commonly used in hydronic space heating applications.

If you're considering this type of system, it pays to investigate any potential mineral, chemical, or bacterial contamination issues before making a final commitment. Run a comparison between the cost of "open-loop" components vs. standard closed-loop components. And perhaps most importantly, verify that local plumbing and mechanical codes will allow this type of installation.

Combined DHW & Space Heating Using Isolating Heat Exchanger :

Given the above concerns it's often preferable to install a heat exchanger between the water heat, and space heating circuits. In this way the space heating circuits remain a closed loop system, isolated from potable water, and thus not subject to mineral or bacterial contamination. Piping components on the space heating side of the heat exchanger can be cast-iron, steel, or other material suitable for closed-loop hydronic systems. Lead-based solders can still be used for space heating piping where needed. Like any other closed-loop hydronic system however, the space heating sub-system requires its own pressure relief valve, expansion tank, air separator, and makeup water system. The heat exchanger is simply replacing what would otherwise be a boiler. Several companies offer compact stainless steel heat exchangers suitable for such applications. A concept drawing for this type of system is shown in Figure 4-51.

Instantaneous Water Heaters:

Instantaneous (a.k.a. "tankless") water heaters have also been adapted for use in radiant panel heating systems. Owing to their intended function of producing hot water from cold water in a single pass, their heating outputs are typically higher than tank-type water heaters. It's important to recognize that many instantaneous water

heaters create relatively high flow resistance in comparison to other components used in hydronic systems. Piping such a unit in series with the other components, (especially radiant panel circuits) could result in very low system flow rates, so low that the flow switch in the instantaneous heater may not even activate the burners/ elements.

The best way to avoid such a problem is to pipe the heater as a "secondary circuit" coupled to the "primary" distribution circuit with closely spaced tees as shown in Figure 4-52. Most instantaneous heaters also contain a flow restricting device based on their intended use water heating application. This device may need to be removed to allow greater flow through the unit when used in a space heating application. Since instantaneous water heaters are turned on by sufficient flow passing through them, system piping must be planned accordingly. When no heat input is needed, the flow through the unit must either stop, or be reduced below the rate at which the flow switch maintains burner op-

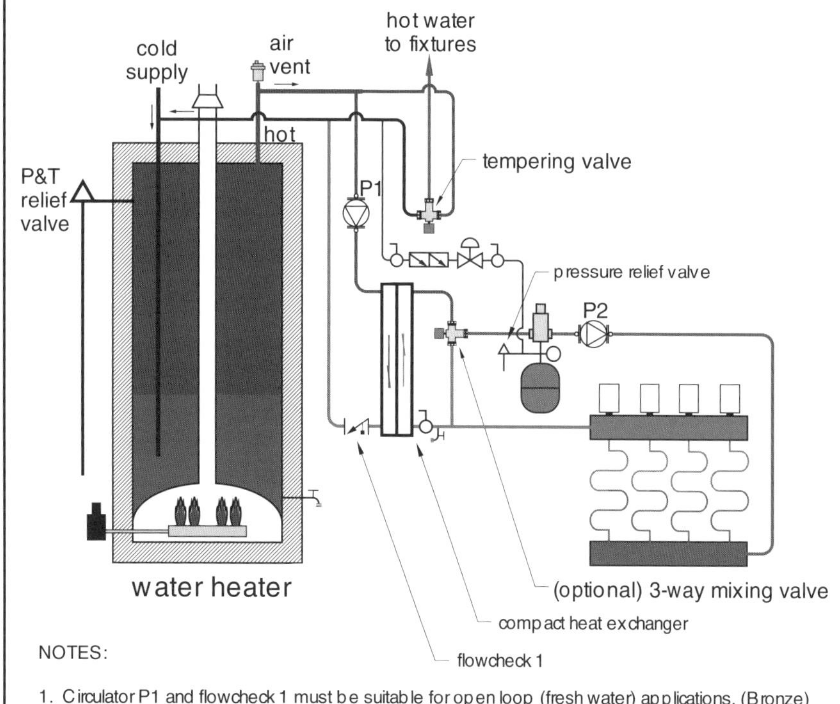

NOTES:

1. Circulator P1 and flowcheck 1 must be suitable for open loop (fresh water) applications. (Bronze)

2. For best performance heat exchanger should be piped counter-flow (as shown).

3. Mixing valve shown in space heating distribution system may not be necessary depending upon operating temperature of water heater and required supply temperature of radiant panel circuits. Other methods of supply water temperature control are possible.

4. Space heating circuit can use cast-iron circulators and other components suitable for closed loop hydronic systems.

5. A pressure relief valve must be provided in the space heat circuit

6. Flowcheck 1 prevents bypass of tank when DHW is drawn. It also prevents gravity circulation.

Figure 4-51 Combination System with Heat Exchanger

eration Again it's advisable to consult the unit's manufacturer regarding any special requirements that may necessary in such an application.

Other Hydronic Heat Sources:

Because many hydronic radiant panel heating systems operate at relatively low water temperature, almost any source of warm water is a potential heat source for them. Space simply doesn't permit detailed discussion of every such possible source, but a short summary paragraph on several of the more common alternatives is in order. Much information has been published on how to properly size and interface most of these devices with hydronic systems.

Active Solar Energy Systems

Although their "heyday" was during the early 80's, active solar energy systems continue to be viable heat sources for radiant panel heating systems. They're usually most economical in areas that have both high solar energy availability, and locally high energy costs.

In nearly all cases they are interfaced with some type of auxiliary heating equipment such as a conventional boiler that can handle the load during extended cloudy weather. The key to achieving good performance is designing the radiant panel system to operate at the lowest possible water temperature. This increases collector efficiency resulting in a higher percentage of solar heating.

Solid Fuel Boilers

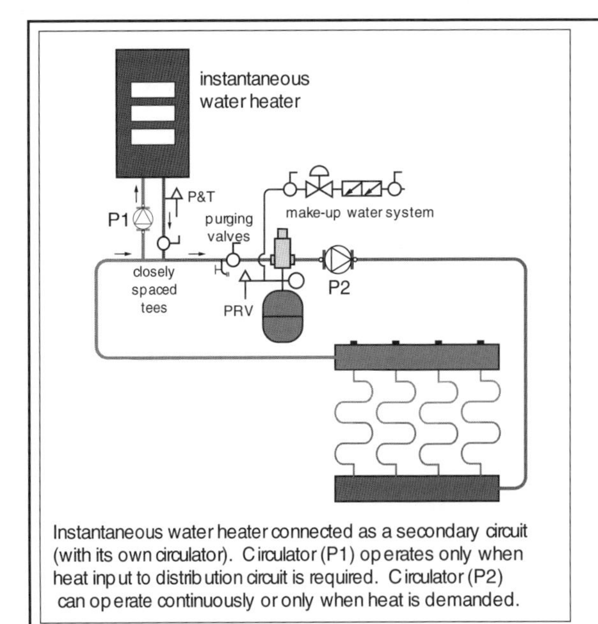

Instantaneous water heater connected as a secondary circuit (with its own circulator). Circulator (P1) operates only when heat input to distribution circuit is required. Circulator (P2) can operate continuously or only when heat is demanded.

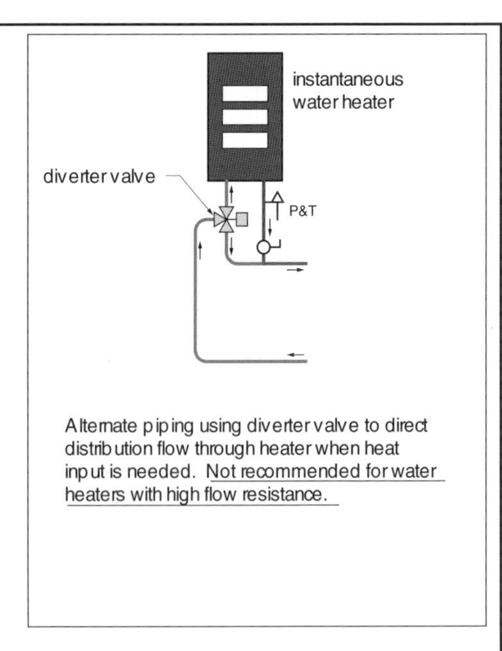

Alternate piping using diverter valve to direct distribution flow through heater when heat input is needed. Not recommended for water heaters with high flow resistance.

- Most instantaneous water heaters require minimum water flow to activate a flow switch which controls burner / heating element operation. Piping design must allow for adequate water flow when heat input is needed.
- Check with water heater manufacturer regarding removal of flow restrictors when unit is used as a dedicated hydronic heat source.

Figure 4-52 Instantaneous Water Heater with Radiant Heating System

There are several types of solid fuel boilers currently on the market, (burning wood, wood pellets, or coal). They range from relatively simple unpressurized wood-burning furnaces, to high-tech wood gasification systems with catalytic converters.

The biggest challenge in applying them has been matching heat output to heat demand during mild weather. Once started, solid fuel boilers cannot simply be turned off like gas and oil-fired boilers. Reducing their output by limiting combustion air produces incomplete combustion which increases both emissions and creosote formation.

One solution has been to install a large insulated thermal storage tank, (like the buffer tank previously discussed for heat pumps, only larger). Water in the tank absorbs heat from the solid-fuel boiler at a high rate, and, if well insulated, stores the heat until needed by one or more building zones. Overall viability again depends on local cost of fuel, as well as a commitment on the part of the owner to keep the unit stoked and cleaned.

Electric Thermal Storage Systems

Many electric utilities across the USA offer time-of-use electrical rates to both residential and commercial customers. During weeknights, (and in some cases all day on weekends and holidays) the unit cost of electricity is substantially lower than peak daytime hours during which it commands a premium price.

By installing a large insulated water storage tank equipped with resistance heating elements, heat energy can be produced during off-peak hours and stored until needed by the radiant panel system. Ideally the heating elements will see little if any operation during peak hours.

Aside from the fact that heat is created with electrical energy, the system essentially gives the same starting point to the radiant panel designer - a large tank full of hot water. Check with local electric utilities regarding availability and pricing of time-of-use rates if interested in pursuing this option.

Air-to-Water Heat Pumps

Several companies make heat pumps designed to extract heat from an air stream blown through the unit's evaporator coil, and then reject the heat at higher temperature to water flowing through the unit's condenser coil. Such devices are commonly used in heat recovery ventilation applications. Like any heat pump, output and efficiency are a function of both air and water temperature. Designing the radiant panel system to operate at the lowest possible water temperature will yield the best performance from the heat pump.

Electric boilers

Boilers that use electric resistance heating elements rather than gas or fuel oil have been available for decades. Several models suitable for residential radiant panel applications are currently on the market. Their viability depends largely on the local cost of electricity, which in many areas of the USA is simply not competitive with natural gas or fuel oil. In areas where their operating cost is justified, electric boilers do have the advantage of not requiring an exhaust system or combustion air supply. Furthermore, no on site fuel storage is required, nor is periodic internal cleaning/ servicing. Most units have several heating elements that can be staged on and off as

required by the load. Short-cycling the elements is also generally not a problem.

Cogeneration Systems

Recent years have seen major growth in cogeneration systems that supply both heat and electricity to larger buildings. This technology is beginning to appear in smaller capacity units suitable for both residential and light commercial buildings. Internal combustion engines operating on either natural gas or diesel fuel drive an electrical generator. Water flowing through the jacket of the internal combustion engine absorbs heat, and carries it to whatever load is waiting. Depending upon the operating schedule of the cogen unit a storage tank may be necessary to interface heat generation with the heat demand of the building. Look for more offerings in this market in the next few years. Radiant panel heating is definitely a viable way to make use of the heat such units produce.

Steam-to-Water Heat Exchangers

Many industrial buildings have process (low-grade) steam available from a central boiler plant. A properly sized heat exchanger can use this steam to create the warm water necessary for a radiant panel heating system. Although application will be limited, this source of heat should not be overlooked when available. Again it has the advantage of not requiring an exhaust system, on-site fuel storage, or combustion air. Temperature control of water going to the radiant panel system is possible by using a thermostatic valve to regulate steam supply.

Electric Radiant Panel Systems

5•1 Introduction

There are several methods of constructing (or installing) electrically-heated radiant panels. They all rely on electric resistance heating, similar to electric baseboard, but offer all the advantages of radiant heat delivery. The radiant panels all operate at relatively low temperature in comparison to other type of electric heat (like a portable quartz heater). These panels can be categorized as follows:

1. Heat Output
   - Constant wattage
   - Self-regulating - Positive Temperature Coefficient (PTC)
2. Voltage applied
   - Line voltage - 100 to 277 VAC
   - Low voltage - up to 50 VAC or VDC
3. Floor installations:
   - Staple-up thin-film panels
   - Staple-up self-regulating PTC elements
   - Embedded cables
   - Embedded cable/mat systems
   - Embedded self-regulating PTC cables
   - Embedded self-regulating PTC elements
4. Ceiling installation:
   - Staple-up thin-film panels
   - Surface mounted panels
5. Wall installation:
   - In-wall low-voltage self-regulating PTC elements.

5•2 Evaluating Electric Useage

The feasibility of using electric radiant heating panels depends on several factors including:

- The local cost of electricity ($/kilowatt-hour), including winter rates and/or preferential rates for total heating
- Charges that may apply based on peak electrical demand (commercial)
- What specific hours of the day the system will be operated
- Whether the system will heat the entire building or only portions of it
- How often the electric radiant panel is used during the heating season

- What the duration is of the panel on-time
- What the feasibility is of using a hydronic versus electric radiant panel
- The amperage capacity of the building's service entrance

It's unfortunate that electric heat in any form is often dismissed from further consideration solely on the basis of the cost of electrically supplied heat versus heat supplied by conventional heating fuels. Although the unit cost of electrically supplied heat is often higher than the unit cost of heat from other fuels, it should not be the sole basis for making such a decision. Electric radiant heating systems offer several unique characteristics that are difficult or impossible to match using other technology. Often times these benefits more than compensate for the higher unit cost of the fuel. Here are a few things to consider (in no particular order of priority):

- Electric radiant panel systems exploit the same advantages of radiant panel heating as do hydronic systems. When properly designed and installed, they will consistently yield lower operating costs in comparison to electric baseboard or electric furnace (forced-air) systems. One study conducted by the US D.O.E. found that electric radiant ceiling panels used 52% less energy than electric baseboard heat.
- Almost every house or apartment has an electrical service capable of supplying at least some electric panel heating. Not every house or apartment has a boiler or even the space to install one.
- In all-electric buildings there's often no need for a natural gas service and its associated monthly minimum service charge.
- Some locations do not have the option of natural gas services and would have to use other services which could prove expensive.
- Many electric radiant panels have very low thermal mass and can respond quickly to changing load requirements when required to do so. This is particularly beneficial in spaces that are heated only when occupied, or have to achieve comfortable conditions on short notice.
- Electric radiant panels are not subject to freeze-up damage.
- Electric radiant panel systems can take advantage of "off-peak" electric rates when available. In some situations it may be feasible to use electric panel heating only during off-peak rate periods, even if another system covers the load during higher rate "on-peak" periods. In some locations off-peak rates apply all day on weekends. Buildings used primarily on weekends may be able to take advantage of such rates using electric radiant panels.
- Electric radiant panel systems are generally less complex than hydronic radiant panel systems. No on-site fuel storage is required. No combustion air source is required as with fossil fuels or wood. No chimney is required. There's no burner to service/clean. There's no possibility of backdrafting combustion gases. And, most installations

5

can be handled by a competent electrician.
- Room-by-room zoning control is simple and inexpensive.
- Properly installed electric radiant panel systems have minimal maintenance requirements and have a record of high reliability.
- Remote locations in buildings are easier to serve electrically than by other means. If a cable can be pulled to the remote location, heat can be provided. Furthermore, the thermal losses of a properly sized electrical cable are insignificant. This is important if the cable has to travel some distance to the radiant panel through cold spaces, underground, or even outdoors.
- The ability of electric radiant panels to service "microloads" or "spot heating" requirements, like a small hobby area in a basement, is exceptional. Although a hydronic system can also serve these loads, it must be carefully designed to prevent short-cycling the heat source.
- Situations that require electric panels to operate with frequent and/or short cycles do not affect the system's efficiency.
- Electric radiant panels also make excellent supplemental heaters for hydronic radiant panel systems.

5•3 Electric Heating Formulas

All electric radiant panel systems rely on the fact that when an electric current passes through any material of known electrical resistance, the amount of heat generated is governed by Ohm's law. The following equations are all "variations" of Ohm's law:

$$W = V \times I \qquad (5\text{-}1a)$$

$$W = \frac{V^2}{R} \qquad (5\text{-}1b)$$

$$W = I^2 \times R \qquad (5\text{-}1c)$$

Formulas 5-1a, 1b, 1c

where:
W= the heat output of the device (in watts)
R = the electrical resistance of the device (in ohms)
V = the voltage applied across the device (in volts)
I = the current flowing through the device (in amps)

Often times it's necessary to convert heat output expressed in watts to Btuh or vice versa. Use the following conversion factor: 1 watt = 3.413 Btuh

5•4 Thin-Film Electric Panels

One type of electric radiant heating panel is manufactured by sandwiching thin carbon "stripes" between layers of polyester film. The carbon stripes, which have a specific electrical resistance determined by the manufacturer, are connected to thin flexible aluminum "bus bars" along each side of the panel. When voltage is applied across the bus bars, a predictable electrical current flows through each of the carbon stripes. Each stripe warms up to dissipate the heat that's generated. The heat con-

5

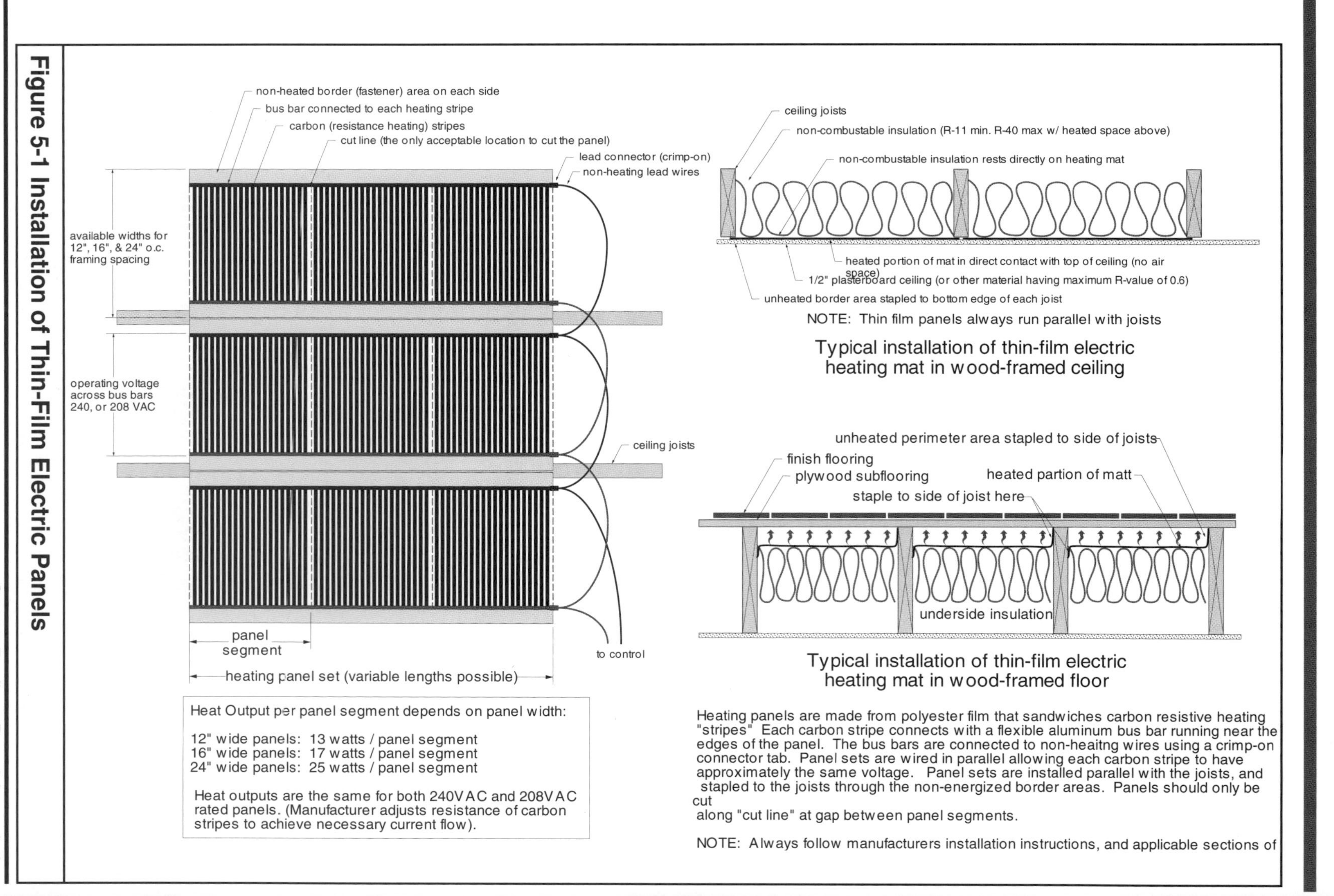

Figure 5-1 Installation of Thin-Film Electric Panels

5

ducts through the thin polyester layer and warms the surface of the panel it is in contact with. In the case of floor heating, the panel releases radiant and convective heat to the air space between it and the underside of the subfloor. Figure 5-1 shows how the thin film panels are constructed and installed.

Thin-film panels are sold in long rolls that consist of many panel segments joined to a common bus bar system. The panels are designed to be installed parallel with floor or ceiling joists. In this way only the unheated border edges of the panels contact the framing. The carbon stripes that generate heat lie centered between the framing. The panels are made for three specific framing spacings: 12", 16", and 24" o.c.

The rolls can be cut to include a given number of panel segments in a given panel set. All cuts must be made along a designated cut line between panel segments. Never make cuts through the carbon stripes as this will alter the heat output properties of the panel segment.

When thin-film panels are installed in ceilings, the panel is unrolled parallel with the ceiling joists and the panel's unheated edges are stapled to the bottom edge of the joists. Panel sets can be of different length as required by the shape of the room. For example, if a room has an angled wall, one panel set may contain 10 panel segments, the one next to it may contain 11, and the one on its other side may contain 9.

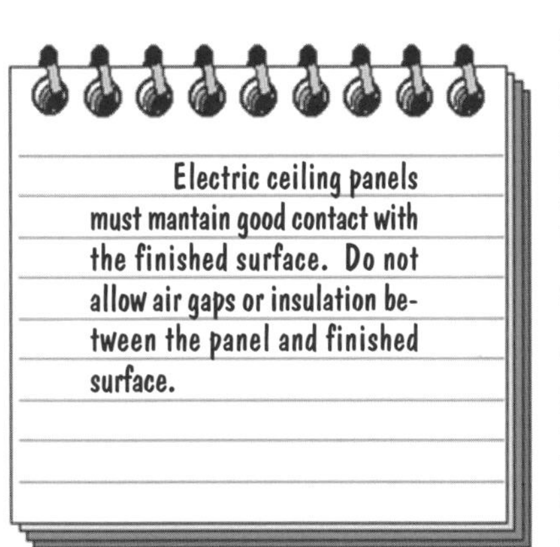

Adjacent panel sets are wired together in parallel to form heating circuits. The number of panel segments that can be combined into a single circuit depends on the heat output of each segment. Always consult the manufacturer's specifications for this information.

Wires that connect adjacent panels do not generate heat themselves. They connect to the bus bars in each panel set using crimp-on connectors supplied by the panel manufacturer. Installers should allow at least 6" of space at the end of each panel segment to provide space to connect these wires.

It is extremely important that thin-film panels maintain good contact with the finished ceiling material. Air gaps, or pieces of insulation between the thin-film panel and ceiling, act as insulators and can force the panel to operate at a higher temperature. It is also important that the R-value of the ceiling material does not exceed 0.6. Higher R-values also force the panels to operate at higher temperatures.

Noncombustible insulation such as fiberglass batts, or blown fiberglass, is installed directly above the panels. As in hydronic systems, this backside insulation is vitally important in directing the panel's heat output in the desired direction. The R-value of insulation for an electric thin-film panel beneath an unheated space can be as high as R-40.

When installed in floors, the non-heated border areas of the panels are bent and stapled to the sides of the joist as shown in Figure 5-1. The surface of the panel runs parallel to, but slightly below, the bottom of the subfloor. Heat flows by radiation and convection from the panel surface to the subfloor. Noncombustible insulation must be installed under the panels to limit downward heat flow. Follow the

5

manufacturer's recommendations.

It's also possible to install thin-film panels at both the top and bottom of framing cavities to heat spaces above and below. Noncombustible insulation must still be installed between the panels as per manufacturer's requirements.

The following is a summary of installation considerations applying to thin-film electric radiant panel systems:

- Follow the manufacturer's instructions explicitly. They are based on National Electrical Code requirements as well as product testing and certifications.
- Be sure panel installation conforms with NEC section 424. Among other factors this code specifies minimum separation distances between the panels and ceiling fixtures or other heat sources like a chimney.
- Never cut panels except along designated cut lines.
- Do not cover ceiling-mounted panel sets with any material(s) that have a combined R-value greater than 0.6 on the heating side of the panel.
- Do not install panels over the top of partitions, cupboards, soffits, ceiling beams, or any other architectural feature that extends up to the ceiling. Again the problem is blocked heat flow and higher panel temperatures.
- Never use any type of combustible insulation (like cellulose or wood chips) over or under thin-film panels. Do not use aluminum foil-faced insulation because of its potential to conduct electricity.
- Never allow any insulation to get between the panel surface and the ceiling surface. Be especially careful of this when using blown fiberglass.
- Only install panel sets parallel with joists. The carbon strips should not be in contact with framing.
- Panel installation to metal joists or furring channels must conform to NEC requirements. All metal framing must be grounded.
- Never install electric panels in areas subject to moisture from roof leaks, snow or rain blown through gable vents, or other sources of moisture.
- Use only controls supplied by the panel manufacturer or having electrical ratings as specified by the manufacturer.
- When completing the job, install the supplied warning labels as specified by the manufacturer. Also provide "as-built" drawings or photo documentation of the installation for future reference.

5•5 Embedded Cable/Mat Systems

Cable Systems

Radiant floors can be created using embedded electric heating cables. A representative illustration is shown in Figure 5-2.

Like thin-film systems, embedded cables rely on a calibrated electrical resistance for predictable heat output. Each foot of cable has a given wattage output rating based on the voltage it's designed to operate at. For example, a cable that outputs 4 watts per foot when operated at 240 VAC will output approximately 3 watts per foot when operated at 208 VAC.

The cables are supplied in various lengths with non-heating "cold leads" spliced to both ends. Each specific length of cable has an electrical resistance set by the manufacturer. Unlike thin-film panels that can be cut in the field, embedded heating cables must NEVER be cut (except at the ends of the non-heating leads). To do so would alter their electrical resistance and affect the output wattage rating. Instead the entire length of the cable must be laid on the floor and embedded. In some cases, it's possible to "use up" extra cable by decreasing cable spacing near the exposed outer wall(s) of a room to increase heat output.

The heat output of the floor is controlled by cable spacing. The closer the cables are spaced, the greater the output from a given floor area. However, adjacent cables should not be spaced closer than 1.5" apart to avoid overheating. Maximum cable spacing also applies to "low profile" systems that have a minimal thickness of embedment material.

Cables are held in place on the subfloor using plastic spacer rails with prespaced notches or by clips (in low-profile applications). The cables do not have to be pulled taut, but should never touch each other or cross over each other. It is important that the splice between the cable and the non-heated leads be buried. The splice is ideally located close to a wall. This allows the non-heated leads to be routed through conduit to an electrical junction

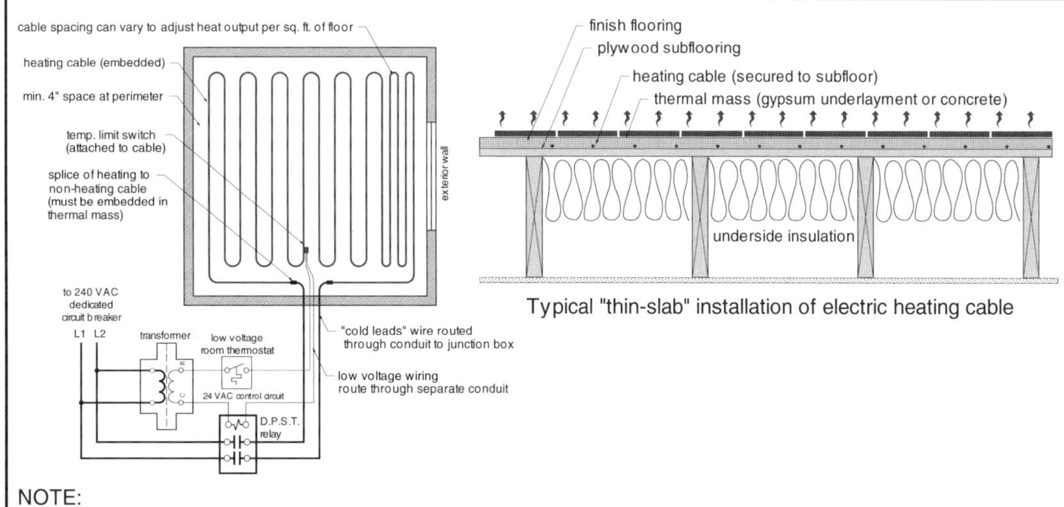

NOTE:

- Cable spacing can be adjusted to increase / decrease heat output in different areas of the room.
- The heating portion of the cable should never be cut. Entire heated length of cable must be embedded in thermal mass.
- Factory-made splices between the heating cable and non-heated "cold leads" must be embedded in thermal mass.
- Different control methods are available. The above schematic is representative only.
- Heating cables should never touch or overlap. Cable should not be spaced closer than 1.5".

Figure 5-2 Embedded Floor Electric Radiant System

5

box.

Several methods of control are possible. Most rely on both room temperature as well as slab temperature. In some cases a standard 24VAC room air temperature thermostat is used as the primary control. Since the cables operate at 240 VAC (or in some commercial buildings at 208 VAC), the thermostat is used to operate a line voltage relay that in turn connects line voltage to the cable. An encapsulated temperature switch is often attached directly to a heated cable and wired in series with the room thermostat. It prevents the cable from operating at temperatures over approximately 140 °F. regardless of room temperature. See Figure 5-2.

The embedded cables can be covered with poured gypsum underlayments or concrete, similar to hydronic thin-slab systems. The same types of finish floor materials used with hydronic systems can be used over electrically heated thin slabs.

Mat Systems

Very thin cable/mat products are also available for floor warming applications. These products consist of a small diameter heating wire woven into a special synthetic mesh. The cable is preformed into a serpentine pattern within the mesh. Cable spacing within the mesh is fixed at approximately 2".

The mesh is unrolled and fit to the room's shape. Non-heating leads are factory-attached to the heating cable at both ends of the roll. The mesh (but NEVER the cable) can be cut, allowing the system to turn corners or go around obstacles in the room. The mesh can be flipped over when necessary to make a corner, but the wires should never touch each other.

Once fitted to the room's shape, the mat is embedded into a layer of thin-set mortar over the subfloor. A second layer of thin-set mortar is applied with a flat trowel over the mesh to fully embed it. Tile, or other hard conductive flooring, is then applied over the thin-set. The entire floor warming layer typically doesn't exceed 1/4" in thickness. Noncombustible underside insulation is required to route heat in the upward direction. The low thermal mass of such a system allows for rapid temperature response. Again it's critical to follow the manufacturer's instructions during installation of these systems.

5•6 Rigid Electric Ceiling Panels

Another type of electric radiant panel consists of a thin layer of graphite-based material with specific electrical resistance properties laminated between polyester layers and mounted into a rigid aluminum frame. The back side of the heating element is insulated with high density fiberglass. The exposed side of the heating element is coated with a textured polymer material, allowing a high percentage of radiant output. The panels can be installed over an existing junction box or fitted with a junction box for retrofit installations.

These panels are available in a variety of sizes from a nominal 2 ft. x 2 ft. up to a nominal 4 ft. x 8 ft. All panels are 1 inch thick. Depending on size, they can operate on either 120 VAC or 240 VAC. They are typically controlled by line voltage thermostats (double pole for 240 VAC panels) on a room-by-room basis. Panel heat output is approximately 50 watts per square foot (171 Btuh/sq. ft.) This is several times higher than the output of a radiant floor or site-built radiant ceiling.

The low thermal mass of these panels allows the radiating surface to reach a temperature of 180 °F to 190 °F in about 4 minutes from when turned on at room temperature. Improved comfort based on increased mean radiant temperature is achieved very quickly. This fast response characteristic is ideal for many applications. Boosting the comfort level in a bathroom prior to taking a shower is a good example. The low thermal mass also prevents residual heat output from overheating a space after the panel is turned off.

Because of the higher surface temperature, these panels should only be mounted to ceilings and should be at least 8 feet above the floor.

Unlike a hydronic system, which is inherently temperature limited by safety controls on the heat source, an electric heating element will always increase its temperature as necessary to dissipate the heat being generated by the electrical conditions under which it operates. For example, the filament of an incandescent light bulb is simply an electric heating element that is "forced" by its design to operate at over 2500 °F.

If necessary, to dissipate the heat being generated, an electric heating element will continue to rise in temperature until it literally destroys itself. The only thing that can intervene to stop an improperly installed heating element from reaching dangerously high temperatures is some type of thermostatic switch.

If, for any reason, that switch cannot detect a high temperature condition because of how or where it's located, the potential exists for the heating element to overheat, possibly to the point of damaging itself or surrounding materials, or even starting a fire. When properly installed, this possibility is all but eliminated. Hence the importance of following the manufacturer's installation instructions cannot be overstressed.

In addition to manufacturer's installation requirements, all electric radiant panels must conform to the National Electric Code (NEC 424-93), or latest revision. Typically the manufacturer's installation specifications reiterate the applicable portions of this code.

5•7 Self-Regulating PTC Cables and Elements

Self-regulating PTC cables and elements use conductive plastic to continually adjust their power output in response to changes in the ambient temperature. This means that when the ambient temperature rises, the electrical resistance increases and the consumption of electricity decreases. Power output is proportional to the voltage squared; small changes in voltage will cause large changes in power output. Also power output varies with geography exterior temperatures, application and insulation used.

PTC cables and elements are useful where even heat distribution is preferred. Good quality PTC cables and elements will not over heat or burn out and there is no need for special controls. Maintaining a continuous even temperature is more efficient and creates less stress on flooring through expansion and contraction than an on-and-off heating system.

5

The PTC cables and elements only draw the energy required to maintain the chosen design temperature. Self-regulating PTC cables and elements are cut-to-length, very flexible and easy to install.

PTC cables and elements are designed for:

1. Total heating in a building
2. Heat-loss replacement for concrete floors
3. Concrete floor warming
4. Tile and stone floor warming
5. Wood and laminate floor warming
6. Resilient floor warming
7. Thermal barrier

PTC Cables

PTC cables are normally powered at 120 to 277 VAC. They do not work on DC voltage. PTC cables can be over lapped and will not overheat. A Ground Fault Circuit Interrupter (GFCI) is strongly recommended.

The cable connects to the branch circuit wiring in a junction box by means of a power connection and end seal. The junction boxes may be distributed around the area to be heated, or collected at a single location.

PTC cable can be installed under a concrete slab or installed in conduit that is embedded in the concrete floor.

When installed in thin-slab application on a subfloor, the cable is held in place using anchoring strips or clips.

Another installation method is to use aluminum plates to help spread the heat over a larger area.

Typical floor coverings can be tile, wood, or laminates. Resilient is not recommended for use over PTC cables attached to aluminum plates, but resilient flooring may be used over slab and thin-slab applications.

PTC Elements

PTC elements operate on low-voltage AC or DC and do not have to be equipped with a GFCI. They are normally connected to a safety-isolating transformer, but can also be run by solar or wind-power source.

Low-voltage PTC elements generate an even temperature over the whole element surface and thereby can be placed closer to the floor surface than any other heating system. With less mass to heat, the system reacts quickly to changes in ambient temperature.

The PTC elements are cut to size on site and are adhered or anchored directly to the subfloor. Refer to manufacturer's recommendation for maximum element length, spacing between strips of element and distance from plumbing fixtures.

The wires can be connected on the floor, up the wall, under the baseboard, or in raised foundation, under the subfloor. The PTC elements must lay flat with no air pockets and be in direct contact with the floor covering. Make the electrical connections following manufacturer's instructions, and always connect the PTC elements in parallel. Minimize voltage drop by planning wire runs as short as possible; follow specifications for low-voltage wire size and length versus load. Select the low-voltage transformer that has the capacity to satisfy the load of PTC elements installed. Maximum load on the transformer is 90% of its total capacity. The transformer must be installed in a well-ventilated area in accordance with the National Electric Code.

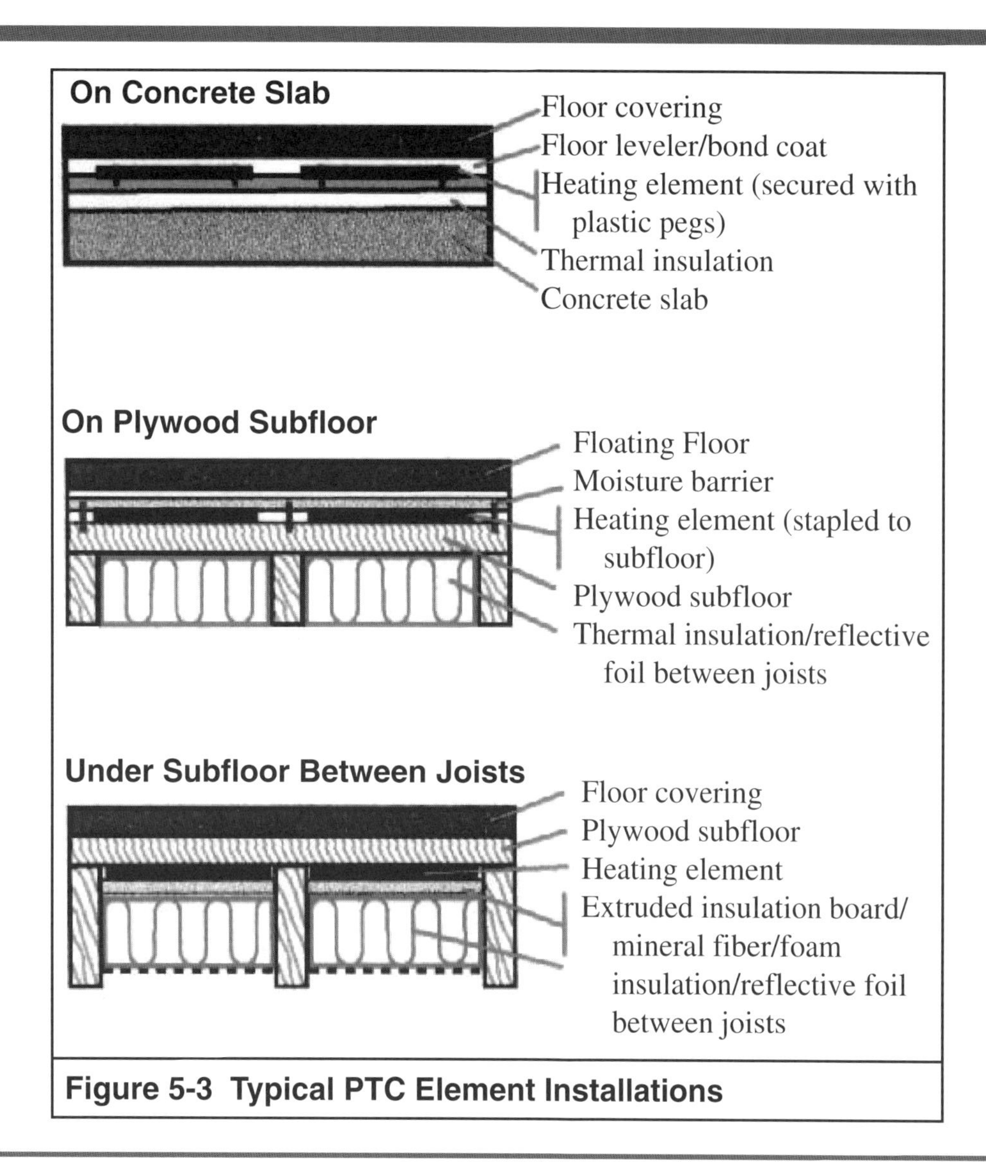

Figure 5-3 Typical PTC Element Installations

5•8 General Precautions

Some general precautions and guidelines for installation of electric radiant heating systems follows:

- Unless otherwise permitted by the NEC and the manufacturer, do not install electric heating elements within 8" of electrical junction boxes. Recessed lights and their trim must be a minimum of 2 inches away from the heating elements. Wiring above heated ceilings must be at least 2 inches above the heating element.
- Do not install extra layers of material to the underside of an electrically heated ceiling. This adds R-value into the path of the heat flow and may force the electric heating elements to operate at higher temperatures.
- Do not install electric heating elements through or over the top of walls, partitions, cupboards, soffits, or other architectural features that

5

extend up to the ceiling. Again the issue is blocked heat flow.

- Never install a suspended ceiling (a.k.a. "drop ceiling") below a heated ceiling. It will block proper heat output to the space below.
- Never cut an electric resistance heating element (cable or mat) unless specifically allowed by its manufacturer. Such cuts will alter the electrical resistance of the element and affect its heat output and operating temperature.
- Make sure the owner understands that nails, screws, or other such fasteners must not be driven into surfaces that contain electric heating elements.
- Only use materials specifically allowed by the manufacturer to cover electric resistance elements.
- Unless specifically allowed by the manufacturer, the heated portion of any electric radiant panel should not make contact with wooden joists, furring strips, or other combustible materials.
- Do not install any type of heating panel in the vicinity of asphalt-saturated roofing felt. Outgassing and odors will occur.
- Never install any electric heating elements in contact with fiberboard, paper-back insulation, cellulose, wood chips, paper, or any other combustible insulation.
- Do not use foil-faced insulation in contact with electric radiant panels. The foil could potentially become electrified in the event of a short or broken wire.
- Do not install electric heating elements where they're subject to moisture from leaking roofs, rain or snow entering through roof vents, or condensation from cold water pipes.
- Use only controls that are supplied by the manufacturer for the system, or, if controls are not supplied, be sure that controls selected meet the specified electrical ratings.
- Never allow heating cables to contact or overlap each other. Observe the minimum spacing requirements set by the manufacturer.
- Any metal joists or furring to which the electrical heating element is attached must be grounded as per the NEC. Verify that the panel being considered is rated for use with metal framing.
- All circuits used to supply electrical heating elements must be "dedicated" circuits. They cannot supply lights, receptacles, or other electrical loads.

Cooling and Air Quality

6•1 Introduction

Many experts feel that, in a perfect world, heat would come from a radiant floor and cooling would come from a separate system above. Since this requires two systems, more often than not the quality and comfort of both systems is sacrificed to combine them into one barely adequate system. Many customers want the comfort of both, they just need it explained to them

Radiant heating does not push air around with a fan as in conventional forced air systems. While radiant heating has many advantages over forced air systems, it does not address basic ventilation or cooling needs required for your project. Many forced air systems also ignore this issue of proper ventilation and just recirculate stale reheated air.

As tighter houses are built, the need for adequate but controlled ventilation has arisen. All homes should have adequate ventilation to assure indoor air quality. Additionally, many climates require cooling. Typically you will be using radiant heat for all the heating needs, but will need cooling for the summers and ventilation to assure air quality throughout the year. This section presents the normal choices and strategies that are available for cooling and ventilating with radiant heat.

6•2 Ventilation and Indoor Air Quality

In climates with mild winters, exhaust only ventilation can make sense since the cost of heating the make up air is minor. Exhaust only systems usually work with a continually operating central fan (sometimes oversized bath fans) and rely on air leakage to let in modest controlled amounts of air to replace the air that is exhausted by the central fan. The fan can additionally be equipped with a programmable timer to only operate when the occupants are home.

Figure 6-1 Kanalflakt Variable-Speed Fan

6

Exhaust only ventilation often uses centralized variable speed fans controlled by timers

Air to air heat exchangers require more duct work but recover heat from the outgoing air and transfer it to the incoming air, thereby saving energy. Air to air heat exchangers require both supply and return ducting, but since they are sized only to provide adequate ventilation and not whole house heating or cooling, it is often possible to use relatively small diameter duct work. In the coldest climates it is also possible to further boost incoming air temperatures with a hydronic coil or duct heater. Heat recovery ventilation can be integrated into air conditioning systems to provide ventilation when there is no cooling load. This saves money on installation costs by eliminating the separate ductwork for a heat recovery ventilator, but adds the complication of controlling the two units which typically have different fan sizes in tandem. More sophisticated air handlers that make this easier are just now being developed. Some heat recovery ventilators can also filter, clean or dehumidify air.

Figure 6-2 Lifebreath Heat Recovery Ventilator

6•3 Evaporative Cooling for Arid Climates

In many arid parts of the western USA evaporative cooling can provide 100% of the cooling needs. Additionally, in many of these climates nighttime temperatures cool off markedly and slab construction is common. The thermal mass of the slab may be cooled at night which will reduce the cooling load during the day. Evaporative cooling systems are inexpensive and standard practice in many arid parts of the country. ASHRAE publishes a chart which list parts of the country that can get 100% of cooling from evaporative means. Since these systems are less costly to operate than air conditioning they should be considered wherever the air is hot and dry. New models that will control humidity by splitting some of the air through a heat exchanger are under development. One detrimental aspect of evaporative coolers is that improperly maintained pads lose efficiency and may develop an odor. Heat recovery ventilation is normally installed separately with evaporative cooling systems since the fan and duct sizing of the evaporative cooler is much larger than normally needed for ventilation only.

Figure 6-3 Inexpensive Evaporative Cooler

6

6•4 High-Velocity Systems

High-velocity systems have been popular for use with radiant floor heating systems because they require little intrusive duct work. They typically utilize 2" structurally reinforced and sound attenuating supply tubing that can be unobtrusively threaded, much like hydronic water pipes, through floors walls and ceilings.

The systems operate at much higher pressures and velocities than conventional air conditioning systems and require special blowers, quality materials and good workmanship for quiet operation. A well installed high velocity system typically has a powerful blower that may consume more energy than a conventional air conditioning blower but may have lower duct losses due to the small easily insulated supply pipes. High velocity systems may run with a lower air temperature to provide more cooling through a small pipe and therefore correct installation to distribute and diffuse the cooled air is important. Heat recovery ventilators may be integrated into the system for ventilation, however due the high-pressure nature of the fan, they must be designed so that both fans can be on and the heat recovery ventilator properly located. They are usually installed separately.

Figure 6-4 High-Velocity Air-Conditioning System

6•5 Ductless Systems

With the compressor mounted outdoors, coolant lines are plumbed to individual condenser/fan units mounted on walls or in ceilings. Often, several fan units can be operated from one outdoor unit allowing zoned cooling. No duct work is required and installation is straight forward. Most units are whisper quiet.

6•6 Conventional Air-Conditioning

Conventional air-conditioning systems are widely understood, frequently and efficiently installed in almost all parts of the USA. Since the radiant system will likely be providing all the heating for the project, the ventilation system can usually be designed for cooling only. This means that the system should be simpler, less expensive to install and work more efficiently.

Ducts often can be smaller and positioned to optimize their cooling benefits without concern for placement that would be required for heating. This system can be designed to provide the ventilation needs of the project by moving modest amounts of air when no air conditioning is required. It can be designed to only run a fan at night when the air temperatures are lower and run the chiller for more cooling only when needed.

Usually, ventilation and cooling systems with radiant are not combined because the duct and fan sizes required are frequently quite different. Ventilation needs are best met by modest amounts of air running continuously while cooling needs usually require bigger fans and ducts, and run intermittently.

Conventional forced-air heating and cooling systems normally have an air handler that attaches to duct work and contains a fan large enough to move the air required for the heating and or cooling system. Typically these air handlers move many times the air required for ventilation alone.

Most traditional forced-air systems do not bring in a percentage of outside air to the air handler to meet the ventilation needs of the project. This has to be done separately as another system or integrated together.

Using the air handler as an integrated part of the heat recovery ventilation system means that the air handler fan may be oversized for ventilation and that the control of the air handler and heat recovery ventilator have to be coordinated. Often it is easier to install them separately assuring the client of a quiet, energy efficient ventilation system.

New air handlers that integrate heat recovery ventilation and have different fan

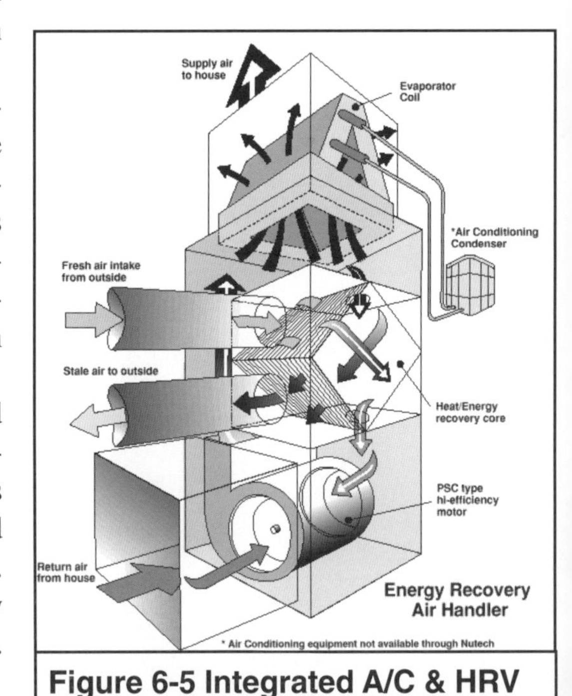

Figure 6-5 Integrated A/C & HRV

speeds for "ventilation only" mode are just now beginning to appear in the market. They should have a good future when combined with radiant since with the addition of an air conditioning coil they meet the two needs most often encountered with radiant; cooling and ventilation for air quality.

6•7 Radiant Cooling

Compromise

It is reasoned that a forced air distribution system can be used to both heat and cool, why not radiant floors? This is logical reasoning and needs to be addressed, but first we should recognize that adding air-conditioning to a forced air heating system requires compromise as well.

In most forced air heating systems the heat is delivered via floor registers. Because warm air rises, floor registers are designed to blow the air horizontally along the floor as far as possible before it rises to the ceiling. Even though this is fairly ineffective with hot air, it works quite well with cold air which would like to stay on the floor anyway. The result is cold air moving across the floor to the cold air return which is normally located low on a central wall. Chilly drafts of very cold air can be unpleasant even when it is 90 °F outside.

The better place, of course, for forced air cooling ducts is the ceiling, dispensing low velocity cool (not cold) air evenly across the ceiling where it gently drops to cold air returns in the floor. This is seldom done.

A similar compromise is made with radiant floor cooling. The cool surface ends up being at the low point where the coolest air is, therefore little natural convection takes place. Close to half the output of a heated floor is from natural convection, cool air coming in contact with the warm floor, rising, giving up heat and falling to be heated once again. Without the assistance of natural convection, floor cooling capacity is about half that of floor heating. As may be obvious at this point, a cooled ceiling has much more cooling capacity than a cooled floor because rising warm air will contact the ceiling, be cooled, fall to the room below where it will absorb heat and rise to be cooled again.

Humidity Issue

Whether a chilled ceiling or a chilled floor is used for space cooling, both systems need to be in a humidity controlled environment. If the system is installed in one of the arid regions of the country, humidity control may not be an issue but in most parts of North America a cooled floor or ceiling will cause moist air to condense on the surface if the surface temperature is below the dew point.

With 70% relative humidity, condensation will occur on a 68 °F surface; 62.2 °F at a 60% relative humidity. This means that some type of air-handling device, which removes humidity, must also be incorporated into the system to keep the relative humidity below 70%.

Added Dimension

Does this mean that radiant panels should not be used for cooling? Not necessarily. Radiant cooling adds a dimension not attainable by forced air cooling systems. It is the same component that is missing in forced air heating... radiant transfer. Since some of the heat we reject is in the

form of radiation, a large cool surface provides a heat sink to draw heat away from our bodies. Heat that can be radiated away reduces the amount that must be convected away and therefore allows us to feel comfortable at a higher air temperature. This, of course, converts into energy savings as well as overall better comfort.

Guidelines

Bjarne W. Olesen, Ph.D. is head of research and development at D.F. Liedelt "Velta" GmbH, in Norderstedt, Germany, president of the European Radiant Floor Association, and an active participant in ASHRAE. In his paper "Possibilities and Limitations of Radiant Floor Cooling" presented at ASHRAE, Dr. Olesen gives the following guidelines for designing radiant floor cooling systems.

• In spaces with seated or standing people, the floor temperature should not be lower than 66 °F for comfort reasons except in areas of higher activity level.

• The heat exchange coefficient between a cooled floor and the room is typically around 1.23 Btu/ft2·hr·°F where 0.97 Btu/ft2·hr·°F is radiant heat transfer.

• Taking into consideration the maximum operative temperature comfort limit of 79 °F for seated individuals, the maximum cooling capacity for a floor system is about 16 Btu/hr·ft2. Dr. Olesen also points out that in spaces with significant direct sunshine on the floor, capacity may double or even triple.

Tube size and spacing can also have an affect on cooling capacity. A floor heating system is often designed with tube spacing of six inches or more. To increase the cooling capacity, it may be necessary to design with closer spacing. Carpet and pad may decrease the cooling capacity by as much as 50%. To avoid supply water temperatures that are too low, supply and return water temperature differences should be kept between 5 to 9 °F.

Where it makes sense

Floor cooling makes the most sense when it is being used in conjunction with an air system. The floor may handle most of the sensible load, while the air system will take care of the latent load. The result is a comfort level not attainable by either system independently. It will also be more energy efficient.

Making choices

While waiting for the floor cooling market to develop and manufacturers to provide packages, it may be wiser to invest in alternative cooling options such as split ductless and high velocity air conditioning systems. Even a simple ducted fan coil can be cost effective. On the other hand, if you like adventure and willing to do some experimentation, radiant cooling could play a significant part in the future of our industry and you may get in on the ground floor. After all, radiant ceiling and floor cooling is the biggest thing to hit Europe since radiant heating.

6•8 Hybrid Solutions

Hybrid radiant heating, forced air heating and air conditioning systems can be designed. Typically some of the home is heated by radiant. The rest of the project might have forced air with air conditioning perhaps throughout. A typical example might be a home where the basement and all areas using a concrete slab have radiant and the rest of the home might have forced air. The heat for the air handler is usually then provided through a coil with hot water from the boiler used in the radiant heating system. If radiant is not preferred by the client everywhere throughout the project, this may be a good option. Hydronics is very versatile and many systems are installed that include multiple forms of hydronic heat, forced air from a fan coil, and cooling with a cooling coil.

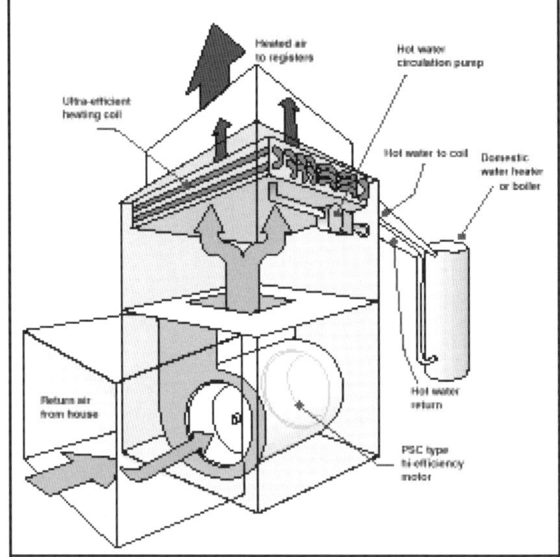

Figure 6-6 Hybrid A/C and Forced-Air Heating

Panel Covering

7•1 Introduction

Floor coverings are an integral part of the radiant panel. The selection of floor coverings can affect the performance of the radiant panel such that the output of the panel may be limited to such a degree that the panel will not meet the full heating load.

Virtually all floor coverings may be used with radiant panel heating by following general selection and installation guidelines.

7•2 Flooring Considerations

Every radiant floor heating system must take the thermal resistance of its floor covering(s) into account. One of the biggest potential pitfalls of radiant floor design is ignoring how much of an effect floor coverings can have on system performance.

The ideal floor covering is no covering at all. Bare concrete slab floors usually offer low thermal resistance to heat flow. Painted, stained, or patterned concrete surfaces all perform essentially the same as a bare slab.

When floor coverings are present, they affect heat output in two ways. First, floor coverings with low R-values allow a greater heat output than do those with higher R-values. Secondly, the R-value of finish flooring affects the variation in surface temperature between the tubes. Higher R-value coverings actually reduce the variation between the highest floor surface temperature directly above the tube and the lowest surface temperature half way between the tubes. In effect, a higher R-value

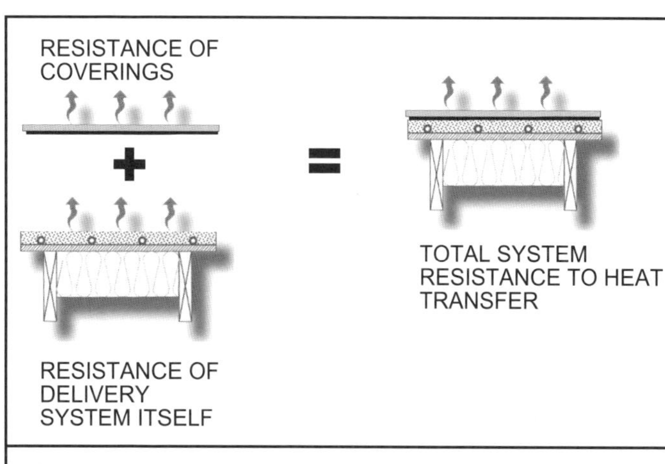

Figure 7-1 Total Resistance to Heat Transfer

7

Material	R-value/in.	Typical Thickness	Typical R-value
Plywood	1.1	3/4"	0.8
Vinyl	1.6	1/8"	0.2
Linoleum Uninsulated	1.6	1/8"	0.2
Linoleum Insulated	1.6	1/4"	0.4
Thinset Mortar	-	1/8"	0.05
Ceramic Tile	1.0	1/4"	0.25
Limestone	1.0	3/4"	0.75
Marble	0.8	1/2"	0.4
Brick	1.0	3/4"	1.5
Wood Flooring Pad	-	1/8"	0.2
Floating Wood Floor	1.0	5/8"	0.625
Laminate Flooring	1.0	1/2"	0.5
Oak	0.8	3/4"	0.6
Ash	1.0	3/4"	0.75
Maple	1.0	3/4"	0.75
Fir	1.2	3/4"	0.9
Pine	1.3	3/4"	1.0
Spruce	1.3	3/4"	1.0
Slab Foam Rubber Pad (33 lb.cu ft)	-	1/4"	0.31
Slab Foam Rubber Pad (33 lb.cu ft)	-	3/8"	0.47
Slab Foam Rubber Pad (33 lb.cu ft)	-	1/2"	0.62
Waffle Rubber Pad (25 lb/cu ft)	-	1/4"	0.62
Waffle Rubber Pad (25 lb/cu ft)	-	3/8"	1.0
Waffle Rubber Pad (25 lb/cu ft)	-	1/2"	1.33
Fiber/Hair/Jute Pad (6-8 lb/cu ft)	-	1/4"	0.97
Fiber/Hair/Jute Pad (6-8 lb/cu ft)	-	3/8"	1.46
Fiber/Hair/Jute Pad (6-8 lb/cu ft)	-	1/2"	1.94
Prime Urethane Pad (2.2 lb/cu ft)	-	1/4"	1.08
Prime Urethane Pad (2.2 lb/cu ft)	-	3/8"	1.62
Prime Urethane Pad (2.2 lb/cu ft)	-	1/2"	2.15
Bonded Urethane Pad (4-8 lb/cu ft)	-	1/4"	1.05
Bonded Urethane Pad (4-8 lb/cu ft)	-	3/8"	1.57
Bonded Urethane Pad (4-8 lb/cu ft)	-	1/2"	2.09
Synthetic Carpet (non wool)	-	1/8"	0.6
Synthetic Carpet (non wool)	-	1/4"	1.0
Synthetic Carpet (non wool)	-	1/2"	1.4
Synthetic Carpet (non wool)	-	3/4"	1.8
Synthetic Carpet (non wool)	-	1"	2.2
Wool Carpet (multiply synthetic values by 1.5)	-	-	-

Figure 7-2 Floor Covering R-values

7

floor covering "forces" the heat to spread out laterally more than a low resistance covering would. This is shown in Figure 7-3.

When selecting finish flooring for heated floor applications, the following factors should be considered:

- Any floor covering system (including finish material and any underlying layers) that has a R-value greater than 2.0 should only be considered when heat output requirements are very low. The term "floor warming" is perhaps more appropriate than floor heating in such cases. Supplemental heating will usually be required.
- Be sure the flooring is warranted by its manufacturer for use over heated floors. Read the "terms" or "limitations" of the warranty as well as any required installation methods. Make sure everyone, including the architect, owner, and general contractor, are aware of any limitations and agree on the flooring to be used.
- Be sure that any adhesives used to bond finish flooring to the heated floor are compatible with the operating temperature of the floor. Most adhesive manufacturers specify the maximum service temperatures for their products. Ask, don't assume.
- Never use any asphalt-based materi-

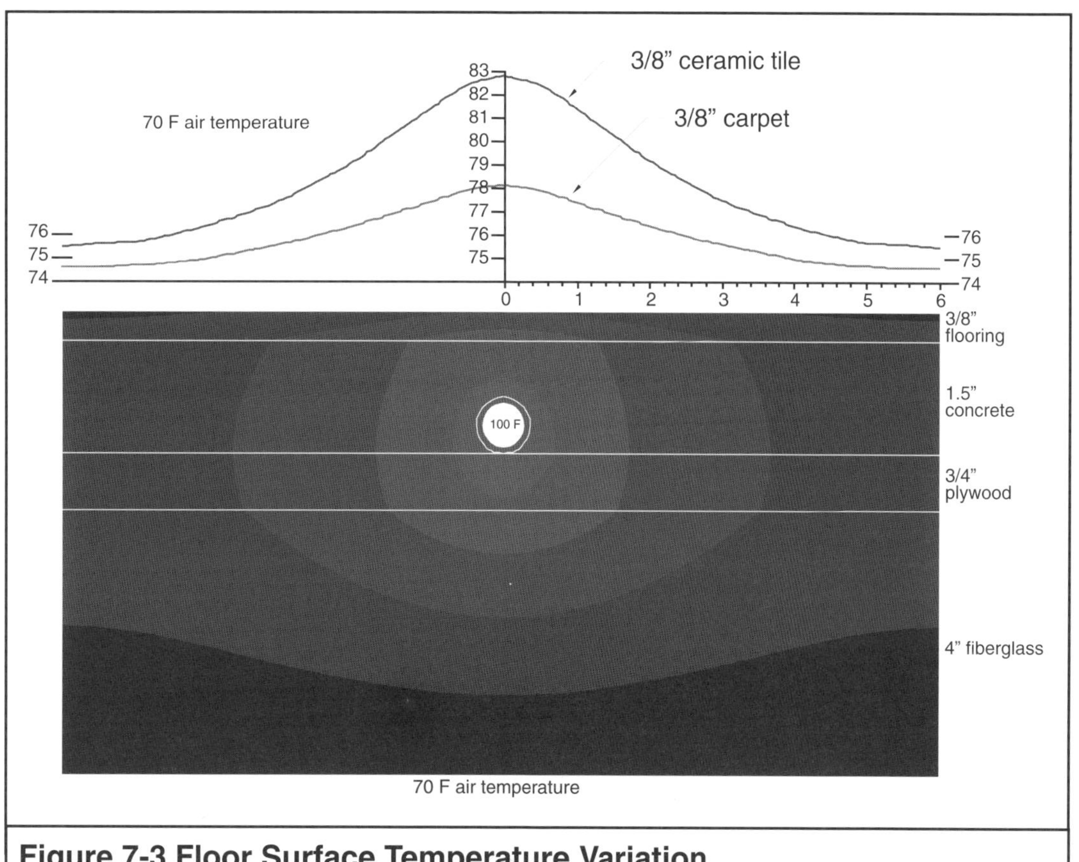

Figure 7-3 Floor Surface Temperature Variation

7

als as part of the floor system. One example would be asphalt saturated roofing felt. When heated, these materials will outgas and create undesirable odors for years.

7•3 Wood Flooring

Many questions and concerns arise when wood flooring is the desired floor covering with radiant floor heating. With proper radiant system design, and proper selection and installation of the wood flooring, many of the concerns can be minimized, if not eliminated. Some things to consider include:

- Laminated hardwood flooring is considered more dimensionally stable over heated floors than is solid-sawn hardwood. Laminated hardwood flooring has veneers with alternating grain directions that reduce lateral shrinkage.
- Quarter-sawn hardwood is considered more stable over heated floors than is "plain sawn" hardwood. Due to its vertical grain direction, quarter-sawn wood experiences less laterally shrinkage.
- Narrow "strip" flooring is less likely to reveal lateral shrinkage than is wider "plank" flooring. Again the concept is to recognize that wood is much more prone to shrinking laterally (perpendicular to its grain lines) than lengthwise.
- All wood flooring should be as dry as possible and temperature acclimated to the interior environment before its installed.
- The substrate onto which the floor is installed should also be tested for dryness. Any heated floors should be operated to ensure their dryness before the wood flooring is installed.
- Hardwood species like oak with a coarse grain structure are less likely to show any minor shrinkage than are fine grained-hardwoods or less stable softwoods.
- All wood floors experience less thermal stress when used over floor heating operated with constant circulation and reset water temperature control. The idea is NOT to create large thermal gradients between the top and bottom of the wood that cause warping. Reset water temperature control attempts to maintain the floor in a quasi-steady state condition, making gradual temperature changes that allow time for the wood to adjust.
- It is absolutely crucial to discuss these flooring issues with owners and architects so that agreement is reached before flooring is selected and installed.

Engineered and Laminate Flooring

Plastic/medium-density (MDF) and/or high-density (HDF) "laminate" flooring has proliferated in recent years with many attractive and easy to clean products. They are thin and have a low R-value that is excellent for radiant heat. This product is not the same as laminated wood flooring, which is made of layers of wood laminated together. The terminology is confusing. Since they are very different products, make sure you know which one is being discussed.

Engineered Flooring

The new laminate floors are typically multiple layers of materials that are thermo-fused together using thin films of glue that bond under heat and pressure. They are usually quite thin and dense which makes them conductive and useful as a floor covering in radiant heating systems. However, since they are a complex composite, some brands are better manufactured than others and more stable under the temperature conditions of radiant.

Figure 7-4 shows a typical engineered floor. The various layers are:

1 - a clear plastic wear layer

2 - the "view" layer

3 - an HDF or MDF board

4 - another HDF or MDF board which has the locking tongue and groove

5 - a plastic bonding layer

It is important that the layers are well bonded and suitable for radiant floor heating. Check with the manufacturer to see if it is warranted for radiant. The instructions for installing laminate flooring over slabs are very specific about moisture content and should be observed.

Prefinished Laminate Flooring

Many laminated wood floors have specific warranties for use over radiant heat, and, since they are cross-laminated with multiple layers of wood and glue, they are dimensionally stable. Most of these systems are edge glued and "float" on the sub floor allowing the floor to expand and contract separate from a radiant thermal mass.

Nail-Down Solid Wood Flooring

The wood flooring installer and the radiant contractor have different concerns that must be reconciled.

A wood flooring installer wants the most stable and a nailable subfloor with good holding power. The wood installer

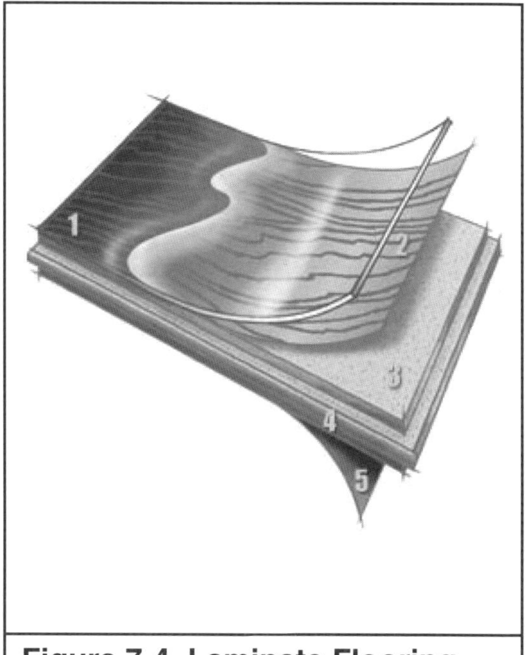

Figure 7-4 Laminate Flooring

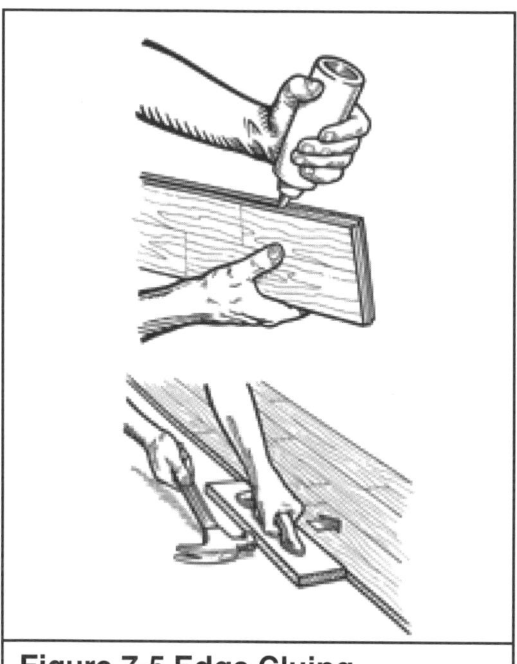

Figure 7-5 Edge Gluing

7

may ask for one or two layers of plywood to nail a traditional wood floor to. However each 3/4" layer of wood adds approximately R-0.8 to the system R-value.

The radiant contractor wants the least number of wood layers to make the heat transfer of the system work correctly. When using a radiant thin slab on a subfloor, direct nailing the wood flooring to the sleepers provides the best heat transfer and is one of the methods in the National Wood Flooring Association Guidelines.

Wood flooring is often successfully installed over radiant, but there are important considerations when doing so. There are multiple wood industry organizations that have guidelines for installing wood floors over radiant heat. Unfortunately there is not complete consensus in the wood industry. Some of the guidelines are more restrictive than others. From a heating system point of view, the wood flooring assembly that offers the least resistance to heat transfer is the most desirable. Some of the recommended assemblies, such as using two layers of plywood for installing wood flooring over radiant, adds significantly to cost and resistance to heat transfer.

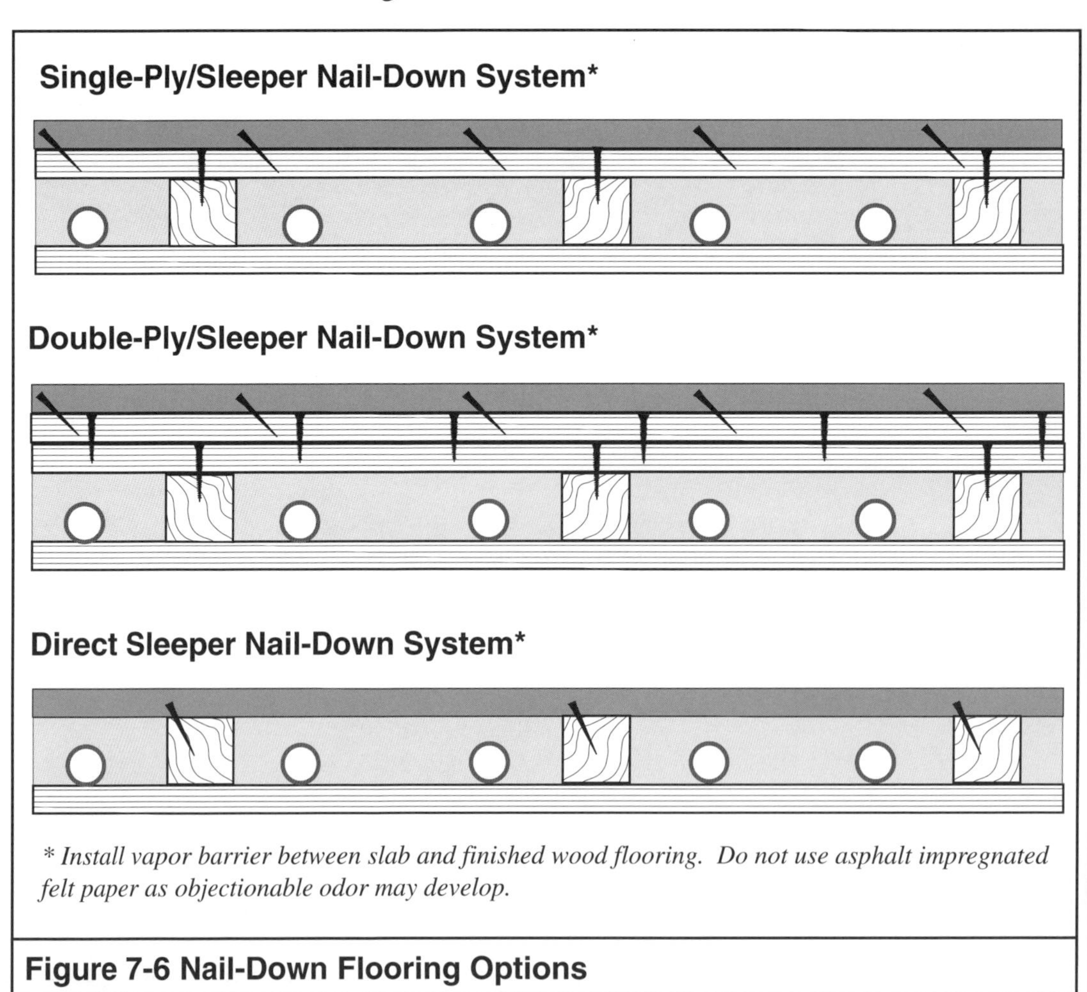

Figure 7-6 Nail-Down Flooring Options

7•4 Ceramic Tile

In general, ceramic tile is an excellent choice over heated floors. It has low R-value, high durability, and usually lasts for decades. The key concern is prevention of cracks. In most cases, cracks in ceramic tile originate in the substrate to which the tile is bonded.

The temperatures of a radiant floor are usually less than what might occur with a dark tile floor on a sunny day behind of a south facing sliding glass door with no curtains. Many tile installations already deal with the temperature range expected from radiant. The rigidity of the floor system, the quality of the adhesive bond, and moisture entrapment issues in glazed tiles with non porous epoxy grout, are frequently of more legitimate concern to tile setters than the temperature changes in a radiant heating system.

The RPA Standard Guidelines ..., Section 15.3: "Ceramic, quarry and marble tile installed over suspended radiant floor panels should have a crack isolation membrane placed between the tile and the thermal mass, or as indicated by tile industry guidelines".

Guidelines in the tile industry do not currently address many of the ways tile is normally installed over radiant heating systems. The National Association Of Tile Contractors Of America (NATCA) has formed a technical subcommittee to develop much more complete guidelines. The Tile Council Of America (TCA) also has guidelines that are being expanded to cover more aspects of radiant heat. These revisions may not be available until later in 2002.

Both organizations realize that there is a natural affinity between warm floors and comfortable tile. Some of the electric radiant floor products have a TCA listing that explicitly describes how the product is to be installed. In the absence of industry guidelines, crack isolation membranes provide a margin of assurance in the installation of tile over many surfaces. Crack isolation membranes were developed to solve problems in all tile installations, not just radiant heating applications.

The membrane is a thin flexible coating or sheet that absorbs the stress of shrinkage cracks in the substrate before they can overstress and crack the tile above. Several such products are currently available on the North American market. Most professional tile installers are familiar with them. It is possible that when the new guidelines are published, some installation methods may not require such a membrane.

The recognized standard for installing all types of ceramic tile is the Handbook for Ceramic Tile Installation published and periodically updated by the Tile Council of America, Clemson, SC, 803-646-4021.

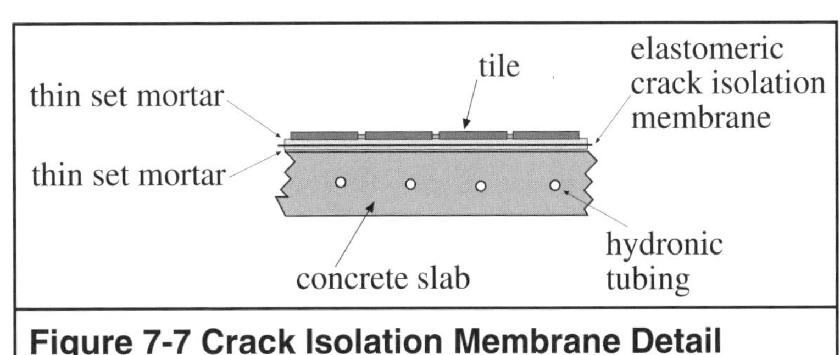

Figure 7-7 Crack Isolation Membrane Detail

7•5 Carpet and Pad

Several types of carpet have been successfully used over heated floors. The key issue again is the R-value of the carpet (or carpet and pad system). A rule of thumb is that the R-value of synthetic carpet is about 2.6 per inch. Hence a 1/4" carpet glued to the floor would have an R-value of about 0.25 x 2.6 = 0.65. This is certainly within an acceptable range for most floor heating applications. Wool carpets have R-values about 50% higher than synthetic carpets of the same thickness and are seldom acceptable over heated floors.

Material	Density	Thickness	R-value
prime urethane	2.2 lb./cu.ft.	1/4"	1.08
		3/8"	1.62
		1/2"	2.15
bonded urethane	4-8 lb./cu.ft.	1/4"	1.05
		3/8"	1.57
		1/2"	2.09
fiber/hair/jute	6-8 lb./cu.ft.	1/4"	0.97
		3/8"	1.46
		1/2"	1.94
waffle rubber	25 lb./cu.ft.	1/4"	0.62
		3/8"	1.00
		1/2"	1.33
slab rubber	33 lb./cu.ft.	1/4"	0.31
		3/8"	0.47
		1/2"	0.62

Figure 7-8 R-value of Carpet Pad

The most limiting component with carpet and pad is the R-value of the pad. Specify that the pad be thin slab foam rubber, thin waffle rubber, or fiber/hair/jute. Prime and bonded urethane pads are not recommended for radiant heating due to their high R-values. For example, 3/8" prime urethane has an R-value of 1.62 while a 1/4" slab rubber pad has an R-value of 0.31. This means that the floor under the urethane must be 40 °F hotter to transfer the same amount of heat than the rubber pad.

Figure 8-8 compares the thermal resistance of some common pads. Higher density (flat) slab foam rubber pads are the best choice for use over heated floors. Their higher density yields lower R-values and provides better support for the carpet, which extends its service life.

Another issue with carpets is outgassing. Heating the carpet will tend to increase any potential outgassing. Carpet or padding with jute or hair backing has been known to cause odors in some floor heating systems. The Carpet and Rug Institute, Dalton, GA, 800-882-8846, has specific standards regarding the outgassing of VOC (volatile organic compounds).

7•6 Resilient Flooring

Resilient flooring is an excellent choice for use with radiant floors since it is a conductive material offering little resistance to heat transfer. Although it has low R-value due to its density and relative thinness, vinyl flooring has a high coefficient of thermal expansion. To avoid wrinkles, vinyl sheet flooring should be adequately stretched when installed and/or bonded to the substrate with an adhesive rated for temperatures at least as high as those it will experience.

Most resilient flooring manufacturers recommend limiting floor temperatures to 85 °F to prevent discoloring. High temperature "striping" caused by poor dispersal of heat from tubing installed without heat transfer plates or thermal mass have been suspected to cause embrittlement in certain older types of vinyl flooring installed over radiant heating systems operating higher than 85 °F.

Two major ASTM standards contain reference to the installation preparations applicable to resilient flooring. ASTM F710 addresses the preparation of concrete floors for resilient flooring goods and contains the 85 °F temperature limitation. ASTM F1482 addresses issues of wooden sub floors, underlayments, and the preparation required for resilient flooring goods. Some manufacturers have a 90 °F temperature limitation on felt backed resilient flooring goods.

Because there are so many different products available, the best advice is to seek competent technical information from the flooring manufacturer regarding installation procedures and warranty limitations.

Figure 7-9 Vinyl Flooring

7•6 Summary

There are hundreds of finish flooring products available. Given the rapid growth in the radiant floor heating industry not all of these products have been tested for suitability over heated floors. The ideal flooring (from a thermal viewpoint) would have relatively low R-value, a low coefficient of expansion, not be subject to thermal degradation, and be fully tested and warranted for use with heated floors. Unfortunately, this doesn't describe the majority of the flooring market at present. However, this situation is improving. Flooring manufacturers are recognizing the growth of interest in radiant floor heating and taking steps to gain (or retain) market share. Look for this trend to continue. In the meantime, don't assume. Use the guidelines discussed in this section and ASK QUESTIONS. Get competent answers from flooring manufacturers backed up in writing if possible. Then be sure all affected parties are advised of your findings.

7

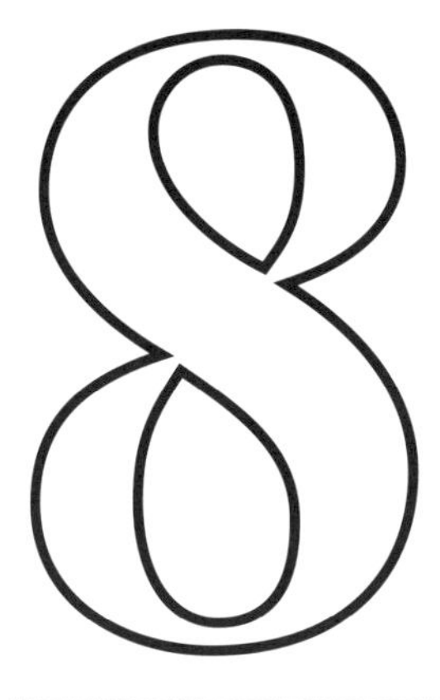

Design Concepts

8•1 Introduction

Throughout the training program you've heard about the many benefits of radiant heating. To deliver these benefits, systems must be properly designed. Previous sections have dealt with qualitative design issues such as tubing installation, system piping, and control methods. This section deals with basic quantitative design. It reviews heating load estimating, then goes on to the concept of heat flux (a key number for describing the performance of nearly all radiant panel systems). It presents a step-by-step calculation procedure for use in sizing a simple floor heating system. The final topic presented is system documentation.

Although it gives some guidance on the subject, this section does not present comprehensive design information for all types of radiant panel systems. There are several publications available through the RPA which present more detailed analytical design information.

Personal computers are now the norm rather than the exception in the offices of most heating professionals. Even the most basic computer system can perform complex design calculations in seconds.

Many manufacturers of radiant panel heating equipment/systems now offer computer software that can dramatically speed up system design. Most of these programs can take the designers all the way from a heat loss calculation right through to a bill of materials. Many can handle both simple and complex systems.

The majority of radiant heating professionals currently use such software on a day-to-day basis. You are encouraged to investigate the various software options currently being offered. As the radiant market continues to grow, look for continuous advances in both the capability and easy of use of these programs.

8•2 Estimating Heating Load

Before attempting to design any kind of heating system, always estimate the heating load of each room in the building using creditable methods and data.

The Design Heating Load of a room is simply an estimate of how fast the room loses heat when the outdoor temperature is at its "design" value. This value is customarily selected as the temperature the outside air is at or above 97.5% of the year.

It's technically called the 97.5% outdoor design dry-bulb temperature, and is published for hundreds of locations across the USA and Canada in the ASHRAE Handbook of Fundamentals. Whenever outside temperatures are above the outdoor design temperature, the rate of heat loss from any room with exposed surfaces will be less than the room's design heating load.

Estimating the design heating load on a room-by-room basis allows the proper size of radiant panel to be selected for each part of the building. The underlying assumption is that the system has to simultaneously maintain comfort in all rooms during design load conditions. Once the design heating load of each room is known, the heat source can be selected based on the sum of these loads.

It should be emphasized that a heating load is a calculated estimate of the rate of heat loss. It is simply not possible to fully account for the hundreds of construction details and thermal imperfections of an object as complex as a building-even a simple house.

Secondly, the heating load is a rate of heat flow from the building to the outside air. It is often misstated, even among heating professionals, as a number of Btus rather than a rate of flow of Btus per hour. Two abbreviations that are commonly used in the USA to indicate rate of heat flow are Btu/hr and Btuh.

When a heating system injects heat into a room at the same rate the room loses heat, the room's air temperature remains constant. By knowing the rate of heat loss from a room we can, at least in theory, replace that heat at the same rate with a radiant panel and thus maintain indoor comfort.

The Mathematics of Heat Loss:

The mathematics required to accurately estimate a room's design heat loss are relatively simple. The key is to stay organized. One thing that helps is to break the room's heat loss up into two categories:

- Conduction heat loss
- Infiltration heat loss

Conduction Heat Loss:

Conduction heat loss is estimated by first identifying the thermal envelope of a room. This envelope includes any room surface that separates heated space from unheated space. Examples include exterior walls, ceilings beneath unheated attics, floors above unheated crawl spaces, as well as windows and exterior doors. Keep in mind that a wall does not necessary have to be an exterior wall to be part of the room's thermal envelope. As long as there is unheated space on the other side of the wall, heat will flow from the warm side to the cool side by conduction.

To calculate conduction heat loss, consider each surface of the thermal envelope to be a "panel" through which heat flows.

Material	R-value *	
Insulations		
Fiberglass batts (standard density	3.17	per inch
Fiberglass batts (high density)	3.5	per inch
Blown fiberglass	2.45	per inch
Blown celulose fiber	3.1 - 3.7	per inch
Foam-in-place urethane	5.6 - 6.3	per inch
Expanded polystyrene (beadboard)	3.85	per inch
Extruded polystyrene	5.4	per inch
Polyicocyanurate (aged)	7.2	per inch
Phenolic foam(aged)	8.3	per inch
Vermiculite	2.1	per inch
Masonry and Concrete		
Concrete	0.10	per inch
8" concrete block	1.11	for stated thickness
w/vermiculite in cores	2.1	for stated thickness
10" concrete block	1.20	for stated thickness
w/vermiculite in cores	2.9	for stated thickness
12" concrete block	1.28	for stated thickness
w/vermiculite in cores	3.7	for stated thickness
Common brick	0.2 - 0.4	per inch
Wood and Wood Panels		
Softwoods	0.9 - 1.1	per inch
Hardwoods	0.8 - 0.94	per inch
Plywood	1.24	per inch
Waferbord or oriented strand board	1.59	per inch
Flooring		
Carpet (1/4" nylon level loop)	1.36	for stated thickness
Carpet (1/2" polyester plush)	1.92	for stated thickness
Polyurethane foam padding (8 lb. density)	4.4	per inch
Slab rubber padding (1/4")	0.31	for stated thickness
Vinyl tile for sheet flooring (nominal 1/8")	0.21	for stated thickness
Ceramic tile	0.6	per inch
Miscellaneous		
Drywall	0.9	per inch
Vinyl clapboard siding	0.61	for all thicknesses
Fiberboard sheathing	2.18	per inch
Building felt (15 lb./100 sq. ft.)	0.16	for stated thickness
Polyolefin housewrap	0	
Poly vapor barriers (6-mill)	0	
Inside Air Films		
Horizontal surface w/upward heat flow (ceiling)	0.61	
Horizontal surface w/downward heat flow (floor)	0.92	
Vertical surface w/horizontal heat flow (wall)	0.68	
45-degree sloped surface w/upward heat flow	0.62	
Outside Air Films		
15 MPH wind on any surface (winter condition)	0.17	

- The data in this table was taken from a number of sources including the ASHRAE Handbood of Fundamentals, and literature from several material suppliers. It represents typical R-values for the various materials. In some cases a range of R-value is stated due to variability of the material. For more extensive data, consult the ASHRAE Handbood of Fundamentals, or contact the manufactuer of a specific produ ct.
- The R-value for a specific thickness of a material may be obtained by multiplying the R-value per inch by the thickness in inches (or fractions of inches).
- The units on R-value are the standard US units of deg.F x hr. x sq. ft. / Btu.

Figure 8-1 R-values of Common Building Materials

8

The total R-value of a wall, ceiling, floor (or any other assembly of materials) is found by adding the R-values of each material that make up the panel, including the R-value of the so-called "air films" that cling to the inside and outside surfaces of the panel. Figure 8-1 lists the R-values of common building materials and air films.

When walls contain windows and/or doors, those areas must be subtracted from the gross area of the wall before the wall's heat loss is calculated. Each window or exterior door would be treated as a separate panel having its own area and R-value.

The conduction heat loss through each panel that forms the room's thermal envelope is estimated. The resulting heat flows are then added to get the total conduction heat loss of the room under design conditions. You can do such calculations on paper, or, even better, with a computer spreadsheet.

Heat Loss Due to Air Infiltration:

Heat loss due to air infiltration is easy to estimate assuming one knows the rate of air leakage of a room. The most common method is based on the number of "air changes" the room experiences per hour. For example, a room with a so-called air change of 0.5 air changes per hour has 1/2 of its heated air volume replaced with cold outside air each hour. The greater the air changes of a room, the greater the associated heat loss.

Unfortunately, an accurate determination of air leakage is difficult. The air leakage rate of a room varies almost constantly. It's affected by the construction materials and the quality of their installation, as well as wind speed and direction, height of the room, and even outdoor temperature. Short of having detailed air leakage studies performed using blower doors or tracer gases, the heating system designer needs to make an educated "guess" as to the air leakage

Building quality	Floor area (sq.ft.)				
	< 900	900-1500	1500-2100	> 2100	add for each fireplace
Best	0.4	0.4	0.3	0.3	0.1
Average	1.2	1.0	0.8	0.7	0.2
Poor	2.2	1.6	1.2	1.0	0.6

Best - buildings with full air barrier wrap, high-quality windows and doors, good weatherstripping/caulking, sealed combustin applicance, and dampers on all exterior air vents.

Average - buildings with plastic vapor barriers, average-quality windows and doors, major penetrations caulked, non-sealed electrical fixture boxes, combustin air drawn from house, and dampered appliance vents.

Poor - buildings with no infiltratin barrier or vapor barriers, no caulking at penetrations or thermal envelope, low-quality (loose) windows and doors, combustion air drawn from house, and no dampers on outside vents.

Figure 8-2 Estimate of Air Infiltration

rate of a room. "Guideline" values are published based on a subjective classification of construction quality and house size in Manual J published by ACCA (Air Conditioning Contractors of America). They are summarized in Figure 8-2.

The total design heating load of a room is the sum of the heat loss due to conduction through all its exposed surfaces added to its heat loss due to air leakage.

Special methods have been developed for estimating the heat loss of basements and slab type floors. Consult the latest edition of ACCA manual J or ASHRAE Handbook of Fundamentals for formula and data.

Special Considerations for Radiant Systems:

The heat loss of rooms is affected in part by the type of heating system used to replace the heat. Many radiant heating systems tend to lower the rate of heat loss from a room for a number of reasons:

- They don't pressurize a room like a forced air system does
- They reduce air temperature stratification which reduces air leakage rates
- They maintain comfort at lower air temperatures than convective heating systems

It has been suggested by some that a conventional heat loss calculation will overestimate the load that must be replaced by the radiant system, and that the size of the heat source in a radiantly heated building can be reduced by as much as 30% over the size calculated by a normal (accurate) heat loss estimate. While some reduction in load is likely in many cases, the decision to reduce the size of the heat source should weigh in the possible consequences of errors in estimating air leakage, the possibility of poorly-installed insulation, or other factors that could easily account for higher losses than anticipated. Every project has to be considered individually. There is, however, strong justification for not oversizing a heat source sized from a normal heat load estimate. In fact, for an all-radiant building there should be a significant "built-in" safety factor when the heat source is sized equal to the total calculated heat loss (with no added safety factor).

It's important to remember that heat loss from a heated floor, wall, or ceiling is affected by the operating temperature of that heated surface rather than room air temperature. For heated floors, wall, or ceilings that are part of the room's thermal envelope, temperature differential components is the temperature difference between the average radiant panel temperature and the unheated air.

8•3 Concept of Heat Flux

All radiant panels operate by spreading out heat delivery over their surface area. If a room is to be adequately heated by such a surface it must be designed to move the room's design heat loss across this area. The ratio of the room's design heat loss (excluding the heat loss of the back side of the radiant panel) divided by the surface area "available" to move the heat across is called the heat flux requirement of the room.

In the case of radiant floor heating, few rooms have their entire floor area available for heat output. Fixed objects like base

cabinets or kitchen islands can cover a portion of the floor area. Installing tubing under such areas is ineffective.

Perhaps less obvious is floor area that is obstructed by movable objects like furniture, area rugs, or commercial inventory. A thick area rug or furniture with skirts that extend down to the floor present a very significant thermal resistance in the path of the heat from the floor. It is crucial to discuss the possibility of such furnishings with clients prior to designing the system and adjust accordingly.

Because furniture and area rugs are likely to be moved, it still makes sense to install tubing under all floor areas other than those covered by fixed objects. A conservative design adjustment would be to discount the estimated area covered by area rugs, furniture, or commercial inventory from the room's floor area when calculating the design heat flux requirement . It's conservative because there still is some heat output from these areas, perhaps 50% of normal, perhaps 25% of normal. It's just not possible to know.

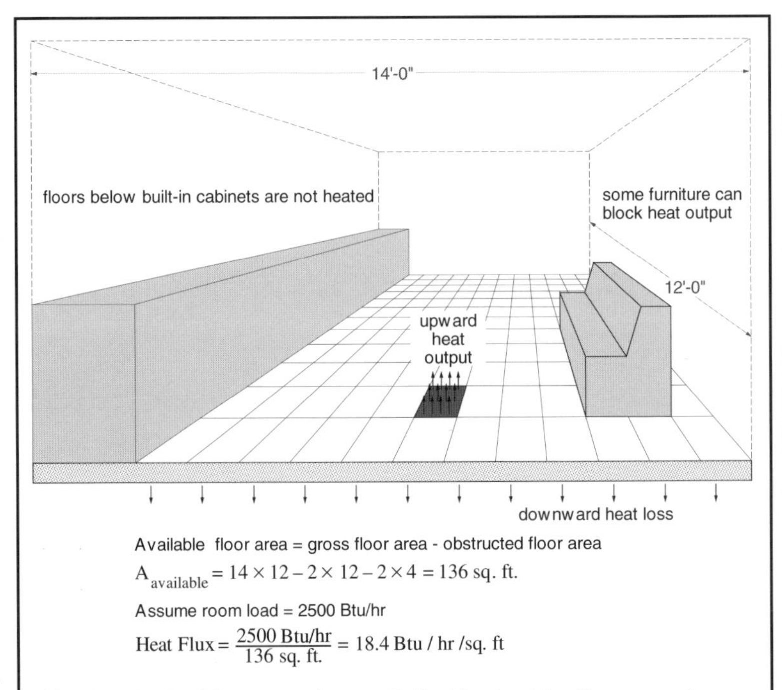

Figure 8-3 Upward Heat Flux Required

8•4 Design Procedure

The following is a simple exercise in system sizing designed to illustrate the elementary steps required in the sizing process. This example is not to be used in actual practice. Use a manufacturers' software or some other proven method for actual job calculations.

Heat Loss Analysis

There are four things you need to know to be able to do a simple heat loss calculation:

- Outdoor design temperature
- Indoor design temperature
- U-value of the partition between the outdoors and indoors

- Partition area in square feet

Use the following procedure, then repeat the same calculation for each wall, ceiling, floor, window and door which is exposed to the outside temperature.

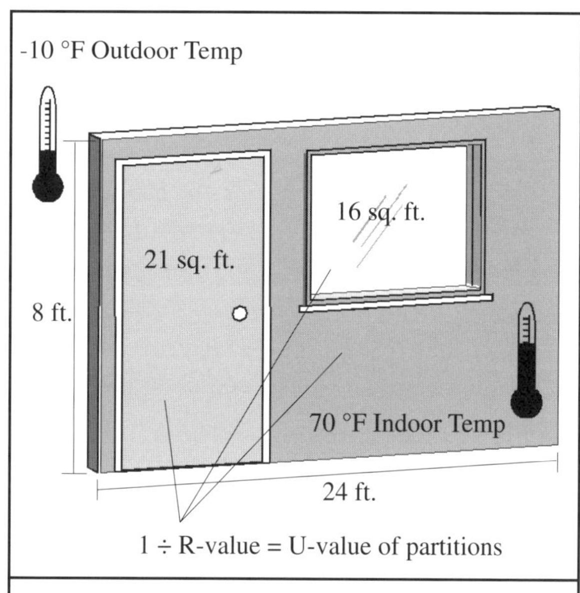

Figure 8-4 Heat Loss Parameters

Step 1

Indoor Design Temp - Outdoor Design Temp = ΔT (temperature difference)

Example: 70 °F minus -10 °F = 80 °F ΔT

Step 2

Partition Height x Partition Width = Area

Example: 8' wall height x 24' wall width = 192 sq. ft.

Note: If the partition is a wall, subtract the area of any windows or doors and figure those separately.

Example: 192 sq. ft. - 16 sq. ft. - 21 sq. ft. = 155 sq. ft. wall area

Step 3

Find the U-value of the partition (1 divided by R-value = U value)

Example: 1 ÷ R-19 = U-0.053

(See Figure 8-1)

Step 4

Multiply ΔT x U-value x Area = Heat Loss (Btuh)

Example: 80 °F x U-0.053 x 155 sq. ft. = 657 Btuh (wall heat loss)

Step 5

Repeat the same calculation for each surface in the room that is exposed to the outdoors and add the results.

Example;

Wall: 80 °F x U-0.053 x 155 sq. ft = 657 Btuh

Window: 80 °F x U-1.06 x 16 sq. ft. = 1,357 Btuh

Door: 80 °F x U-0.486 x 21 sq. ft. = 816 Btuh

Ceiling: 80 °F x U-0.050 x 480 sq. ft. = 1,920 Btuh

Total Surfaces Heat Loss = 4,750 Btuh

Note: Be sure to add to the heat loss ceilings that are exposed to attics and floors that are over crawl spaces. Do not add floor slabs which are poured on or below grade.

Step 6

Calculate the Air Infiltration Loss and add it to the Total Surfaces Heat Loss

Multiply the room Length x Width x Height x ΔT x Air Changes per Hour x 0.018

Example:

24 ft x 12 ft x 8 ft x 80 °F x ACH-0.5 x 0.018 = 1,659 Btuh

Step 7

Determine the Total Heat Loss by adding the Total Surfaces Heat Loss and Air Infiltration Loss together.

Example:
4,750 + 1,659 = 6,409 Btuh

Floor Output Required

Step 8

Determine the Available Floor Area by subtracting any area that will not receive radiant heat such as under cabinets to get the Adjusted Floor Area.

Length x Width - Non Heated Area = Adjusted Floor Area

Example:
12 ft x 24 ft - (2 ft x 10 ft) = 268 sq. ft.

Step 9

Determine Floor Output Needed by dividing the Total Heat Loss by Adjusted Floor Area.

Example:
6,409 Btuh ÷ 268 sq. ft. = 24 Btuh/sq. ft.

Required Floor Surface Temperature

Step 10

Consult Figure 8-7 to find the Required Floor Surface Temperature for the desired room temperature. Note: as the room temperature changes, the floor surface temperature changes respectively.

Establish a maximum acceptable floor temperature for each room. Generally a maximum floor surface temperature of 85 °F is used for rooms with prolonged foot contact. However, this temperature may be as high as 92 °F. in transition areas like entry foyers or hallways.

		Room Temperature					
		60 °F	65 °F	68 °F	70 °F	72 °F	75 °F
Btuh/sq. ft.	10	65.0	70.0	73.0	75.0	77.0	80.0
	15	67.5	72.5	75.5	77.5	79.5	82.5
	20	70.0	75.0	78.0	80.0	82.0	85.0
	25	72.5	77.5	80.5	82.5	84.5	87.5
	30	75.0	80.0	83.0	85.0	87.0	90.0
	35	77.5	82.5	85.5	87.5	89.5	92.5
	40	80.0	85.0	88.0	90.0	92.0	95.0

Figure 8-5 Required Floor Surface Temperature

Tube Spacing and Circuits Required

Step 11

Look up the suggested tube spacing (for each room) based on its floor output requirement using Figure 8-6. These tube spacings are "conservative" and are selected to limit variations in floor surface temperature to acceptable ranges.

Example:
3/8" tube @ 24 Btuh/sq. ft. = 6" o.c.

Req. upward Btuh/sq.ft.	3/8" I.D.	1/2" I.D.	3/4" I.D.
<10	9"	15"	18"
10-20	9"	12"	15"
20-30	6"	9"	12"
30-40	4"	6"	9"
Rec. Length	200'	300'	500'

Figure 8-6 Tube Spacing: Inches on Center

Step 12

Locate the spacing in Figure 8-7 to find the appropriate spacing factor. Multiply the Available Floor Area by the Spacing Factor to determine the Tubing Required for the room.

Example:
268 sq. ft. x 2.00 = 536 ft. of tube required

Spacing	Factor
4"	3.00
6"	2.00
9"	1.33
12"	1.00
15"	0.80
18"	0.67

Figure 8-7 Spacing Factor

Step 13

Divide the Tubing Required by the Recommended Length from Figure 8-6 to determine the Number of Circuits Required.

Example:
536 ft. ÷ 200 ft = 2.7 circuits required.
Round up to 3.0 circuits

Step 14

Divide the Tubing Required by the Number of Circuits to determine the Length per Circuit.

Example:
536 ft. ÷ 3 = 179 ft. per circuit

Supply Water Temperature

Step 15

Floor coverings will have a major impact on the supply water temperature. High R-value floor coverings will require that the supply water temperature be raised to overcome the resistance to heat flow through the floor covering. A highly conductive floor covering such as ceramic tile will require a lower supply water temperature to provide the same heat output compared to carpet and pad. Use Figure 8-8 to select a Supply Water Temperature based on the floor covering R-value and heat output required.

Determine Flow Rates and Head Loss

Step 16

To select circulating pumps for the system, it is necessary to determine the system flow rate, circuit flow rate, and circuit head loss. A good "rule-of-thumb" for determining flow rate is to divide the total heat loss by 10,000. The number 10,000 represents the approximate Btuh delivered with 1 gallon per minute flow and a 20 degree temperature drop between supply and return.

Example:
6409 Btuh ÷ 10,000 = 0.64 GPM

R-value	Floor Covering	Buth/sq.ft. 15	20	25	30
Tube on subfloor					
0.2	vinyl/tile	85 °F	90 °F	94 °F	98 °F
0.5	3/4" hardwood	90 °F	110 °F	116 °F	123 °F
2.2	1/2" carpet with 1/4" rubber	120 °F	133 °F	148 °F	155 °F
3.8	1/2" carpet with 1/2" urethane	150 °F	-	-	-
Tube below subfloor					
1.2	vinyl/tile	105 °F	112 °F	120 °F	130 °F
1.5	3/4" hardwood	110 °F	130 °F	136 °F	143 °F
3.3	1/2" carpet with 1/4" rubber	140 °F	153 °F	168 °F	175 °F
3.8	1/2" carpet with 1/2" urethane	170 °F	-	-	-

Figure 8-8 Supply Water Temperature (based on 70 °F indoor temperature and 20 °F Δt across supply and return)
Note: Temperatures are approximate and for illustration purposes only.)

Step 17

Then, divide the system GPM by the number of circuits to determine the flow per circuit.

System GPM ÷ # Circuits = GPM per Circuit

Example:
0.64 GPM ÷ 3 = 0.213 GPM/Circuit

Step 18

Find the head loss for the tube size and flow rate from Figure 8-9.

Example:
0.213 GPM/Circuit for 3/8" tubing = 1.41 Ft H_2O per 100 ft.
1.41 Ft H_2O x circuit of 179 ft. ÷ 100 ft. = 2.52 Ft H_2O in circuit

GPM	Ft H_2O per 100 Ft Tube		
	3/8"	1/2"	3/4"
0.1	0.29	0.07	0.01
0.5	4.78	1.12	0.22
1.0	16.06	3.77	0.73
2.0	54.02	12.67	2.45

Figure 8-9 Head Loss per 100 Ft. Tube at 140 °F
Note: Head loss numbers are approximate and for illustration purposes only.)

Step 19

This sample 3-circuit zone will require a pump capable of providing 0.64 GPM at 2.52 Ft H_2O for the tubing only. Additional head loss will occur in system piping, fittings and manifolds. Determine all component heat losses and add them together to determine the total head loss required.

8•5 System Documentation

Most radiant panel systems are designed to last for decades. Over their life all are likely to need occasional servicing. Some may be expanded or modified as the buildings they serve are altered. The same heating professional who installed the system may not be available to service it over its entire life. All these situations present potential problems for systems that are not adequately documented.

Benefits of Documenting Your Systems:

Good system documentation pays for itself in several ways. Here are a few benefits to consider:

- It helps ensure the system will be installed as the designer intended
- It establishes a permanent technical description of system (both on and off site)
- It allows faster troubleshooting by new service technicians
- It demonstrates a professional attitude to prospective clients
- It helps ensure a long service life over multiple generations of replacement components and service personnel
- It speeds design/drawing of future systems

Types of Documentation:

The following materials (collectively) are suggested as a means of documenting a hydronic radiant panel system:

- Piping schematic of entire system
- Electrical control schematic of the system
- Scaled floor plans showing placement of radiant tubing and other heat emitters

8

- Description of system operation
- Component data sheet manual
- Photographs and/or videotape of "as built" installation
- Service record

You've seen many examples of generic piping schematics in this manual. However, the documentation drawings for your systems should be as specific as possible. Equipment should be called out by manufacturer and model number. Electrically-operated components like pumps and zone valves should get designations such as (P2) or (ZV3). These designations should appear consistently in both piping and electrical schematics. They allow the installer or troubleshooter to identify the component on either drawing and then find its function and/or location on the other. Ideally the hardware itself should be tagged or labeled with the same designations.

Figure 8-10 shows a representative piping schematic for a multi-load hydronic system supplying radiant floor heating, hydronic baseboard, and domestic hot water. Note the drawing contains a symbol legend for those who may not be accustomed to the symbols you use.

Figure 8-11 shows a wiring diagram for the system whose piping is shown in Figure 8-10. This diagram uses a ladder diagram format to clearly show both line voltage and low voltage portions of the control system. Again a legend is shown to assist those unfamiliar with the schematic symbols. Note how the component designations correspond between Figure 8-10 and Figure 8-11.

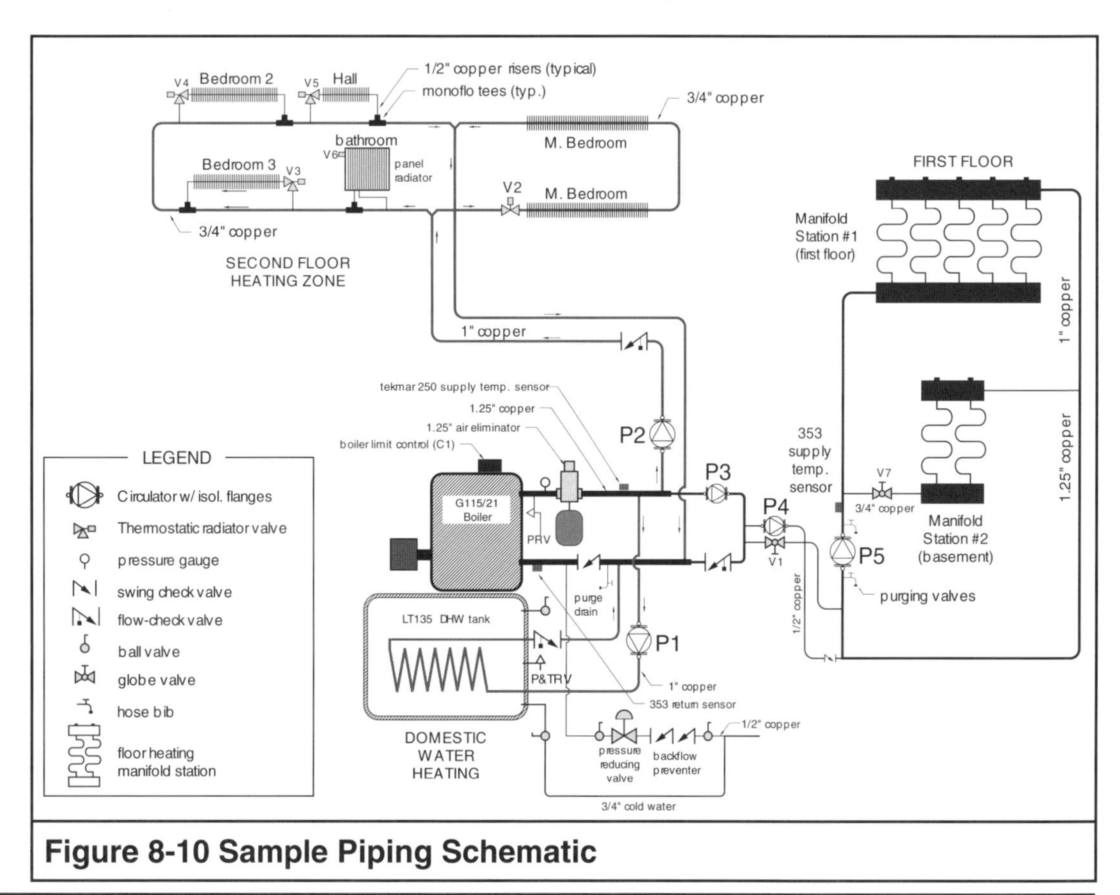

Figure 8-10 Sample Piping Schematic

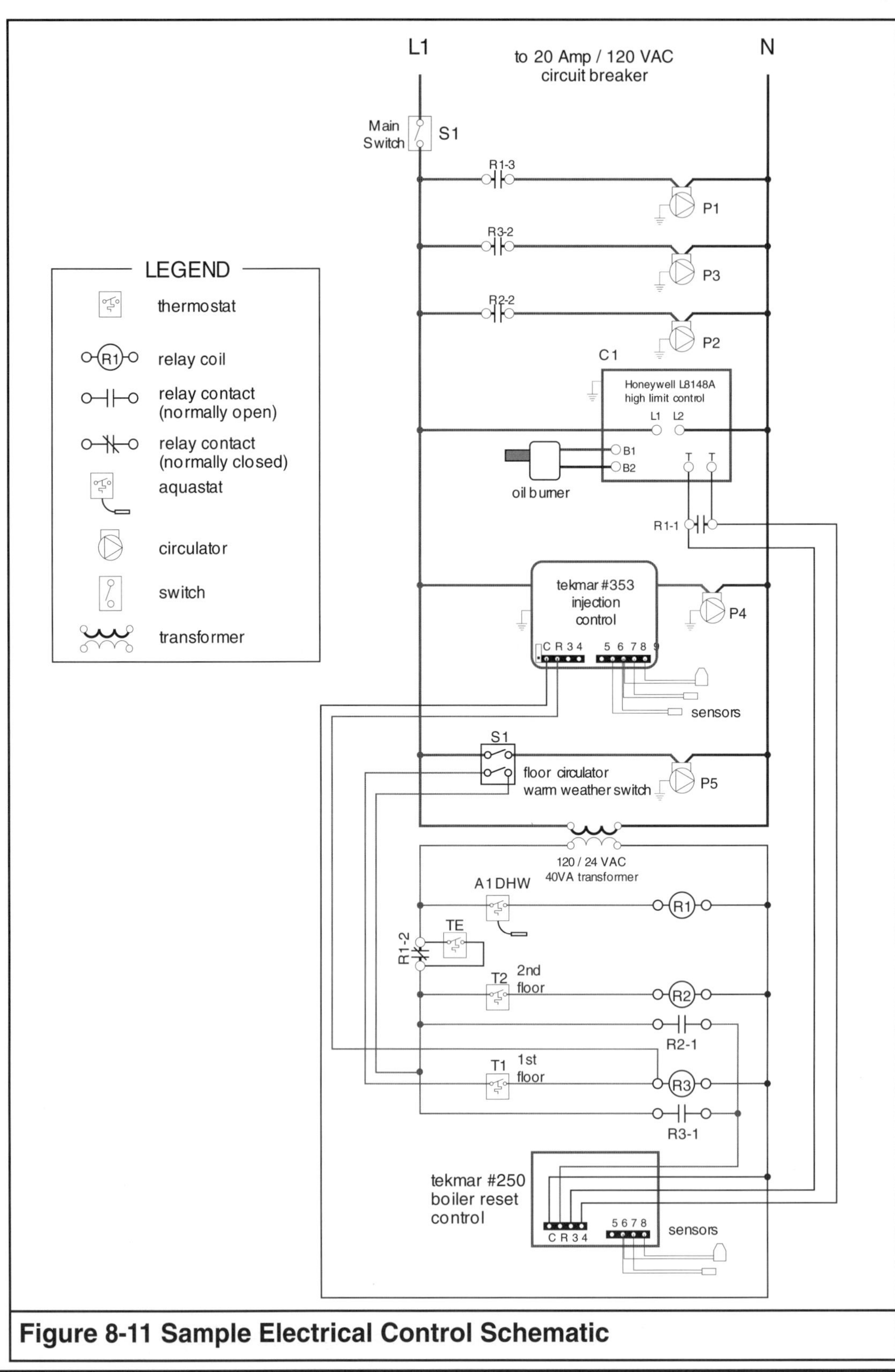

Figure 8-11 Sample Electrical Control Schematic

Tubing Layout Drawing:

Producing an accurate tubing (or cable) layout plan for every radiant panel job has many benefits. These include:

- Providing the installer with a detailed plan of how the tubing should be installed. This avoids misunderstandings, "trial and error" installation methods, incorrect coil cuts, or other results that differ from those expected by the designer.
- Much faster installation is possible with an accurately drawn plan. In many cases the time spent drawing the tubing layout plan is more than recovered by reduced installation time, especially for new installers.
- Circuit lengths can be accurately determined before tubing is purchased or installed. This avoids excessive circuit length, as well as finding that the tubing segment being installed is too short to make it back to the manifold after the majority of it has been fastened in place. It also allows more detailed quantity estimating, as well as "optimization" of how individual circuits should be cut from both existing and new coils.
- The plan provides a permanent/ accurate record of where the circuits are installed, and what manifold station they are connected to. This can save significant time and frustration during balancing, service, or when building renovation is performed.
- It provides other trades on the job (both present and future) with information that can prevent damage to embedded tubing circuits

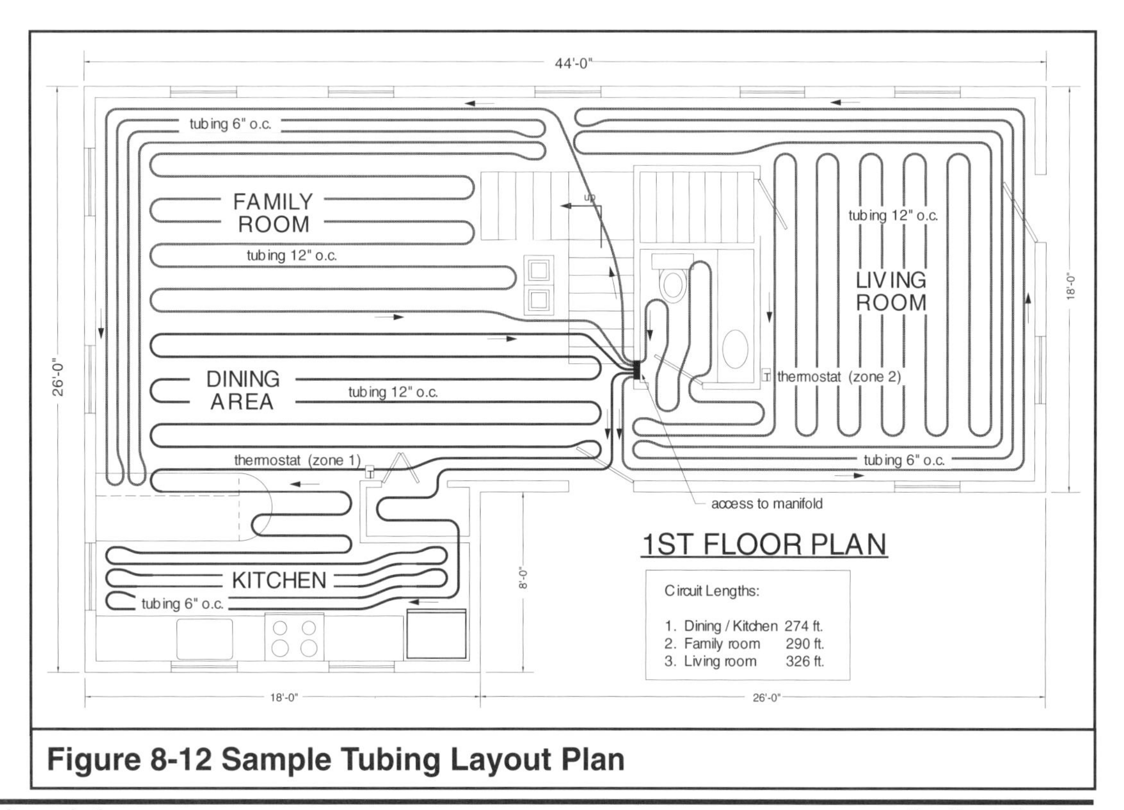

Figure 8-12 Sample Tubing Layout Plan

from "blind" drilling, sawing, etc. In nearly every case accidental damage to embedded tubing occurs because someone didn't realize it was there.

Although it's possible to manually draw a tubing layout plan, the availability and relatively low cost of computer-aided-drawing (CAD) programs make this the preferred method of producing the plan. Several CAD programs costing less than $300 are presently available for both Windows® and Macintosh® computers. All are capable of precision scale drawing, and most can also be configured to automatically determine the length of completed circuits. When the plan is completed, it can be printed to any number of low-cost ink-jet or laser printers.

Optimal drawing techniques will vary from one CAD program to another. Before deciding on a particular program the designer should contact the software vendor and describe exactly what it is that the program needs to do. The vendor should then be able to recommend a specific drawing technique for their software.

Most drawings would begin by establishing an accurate 'template" of the building's floor plan. Although this template needs to be drawn to accurate scale (to allow the program to accurately measure circuit lengths) it does not necessarily have to include all the details shown on a typical floor plan. Remember that the drawing's purpose is for placing tubing, not for locating windows, doors, receptacles, or other architectural elements.

In cases where the building plans have also been drawn using a CAD system, the building designer may be able to provide you with a disc file containing wall and partition locations. Many of the current CAD programs can import such a file in various formats such as DXF or DWG. If such files are available they can often save time in constructing the tubing layout plan, especially for complicated floor plans. It's advisable to "strip out" any irrelevant details, notes, dimensions, etc. from such CAD files before laying out the tubing circuits since they only tend to clutter the plan. When the basic floor plan template has been established it is also advisable to draw each floor circuit on a separate "layer" in the CAD program. This allows easy modifications, deletions, etc. without disturbing other elements of the plan.

The following items are suggested for inclusion in the tubing layout plan:

- Name of project and the scale at which the drawing is printed
- Name of each room
- Exact location of each manifold station within the building (with placement dimensions)
- A name for each manifold station
- The direction from which each manifold station is accessed once walls are closed in
- The routing path of each circuit
- The flow direction in each circuit
- A name for each circuit for reference in a list (if other than room name)
- The location of all control/construction/expansion joints in slab
- A schedule listing each circuit by name and length
- The location of all thermostats used
- The location of any valves or controls other than those at the manifold station(s)
- All locations where floor is to be

marked to prevent accidental damage to tubing
- Notes and/or details relevant to the installation (if not provided on other drawings)

Figure 8-14 is an example of a tubing layout plan for a hydronic radiant floor heating system using a low-cost CAD system.

Description of Operation:

Good system documentation should include a written "description of operation" (Figure 8-15) along with the drawings mentioned above. Describe how the system is composed of sub-systems such as domestic water heating, floor heating, garage heating, etc. Then describe how each subsystem is intended to operate. Make frequent referral to components shown in both the piping and electrical schematics using their abbreviations (i.e. Upon a call for heat from thermostat T2, boiler firing is enabled through the boiler limit control C1, and circulator P2 is turned on). Such descriptions reveal the sequential and/or priority nature of many control systems, something that's not obvious from the schematic drawings.

Components Data Sheets Manual:

Many of the components used to assemble a radiant panel heating system come with installation instruction and/or specification sheets. It's a good idea to keep a file for all such sheets for future reference. Keep the file handy on the job site while the system is being installed. Then transfer all the sheets into a 3-ring binder. Label the binder with the name of the job and leave it in a safe place in the mechanical room for future reference.

Photo Documentation:

Documentation is not limited to paper. Given the inevitable variations between plans and "as built" construction, photographing or video taping critical areas of the installation is cheap insurance against a short memory. On floor heating jobs photograph the area around each manifold station before the slab is poured to allow circuits to be quickly matched with their associated manifold valves in case proper manifold labeling is missing. It's also good to photograph areas where future penetrations of the slab might occur, such as where equipment will be fastened to the floor. These photos are sometimes more useful when a tape rule (with large numerals) is laid across the tubing with its end up against an easy to relocate reference such as a wall or a column. Make sure the numbers on the tape can be read before snapping the photo. And when the photos arrive, take a few minutes to write the project name, date, and any specific details on the back before you file them.

Project Service Record:

A final part of documentation is to provide an accurate service record of the system. Whenever the system is serviced or adjusted the specifics should be noted and dated in this manual. This information shows a new service person the history of the system and may save hours of frustration in tracing down a malfunction.

DESCRIPTION OF SYSTEM OPERATION:

1. DOMESTIC WATER HEATING: Upon a call for water heating from tank aquastat (A1), boiler firing is enabled under control of the boiler limit control (C1) activated through relay contact (R1-1). Boiler temperature will climb toward the upper limit setting and, after reaching the high limit setting, cycle on/off based on limit control's differential. Circulator (P1) is turned on through relay contact (R1-3) to circulate hot boiler water through the heat exchanger in the DHW tank.

Domestic water heating is the priority load. When domestic water is being heated, all other circulators, except the floor circulator (P5) are temporarily shut off by interrupting control voltage through relay contact (R1-2).

FAIL-SAFE PRIORITY OVERRIDE: Should the system fail to come out of priority domestic water heating mode due to a control failure, room thermostat (TE) will detect an abnormally low room temperature and then bypass the open priority contacts (R1-2), allowing space heating to resume on both 1st and 2nd floors.

2. FLOOR HEATING (1st FLOOR): Circulator (P5) is turned on at the beginning of the heating season through the manual switch (S1). It maintains continuous circulation through the floor circuits during the heating season, and is shut off at the end of the heating season.

Switch (S1) also enables a 24 VAC circuit through room thermostat (T1). Upon a call for heat from thermostat (T1), boiler firing is enabled under the control of the tekmar #250 boiler reset control (see suggested settings below) which is powered up through relay contact (R3-1). Circulator (P3) is turned on through relay contact (R3-2) to transport boiler water to the injection mixing piping, and provide bypass water to boost boiler return temperature. Injection mixing is enabled when 24 VAC is supplied to tekmar control #353 through thermostat (T1). This control operates circulator (P4) at a speed suitable to achieve the necessary temperature to the manifolds. See instructions supplied with tekmar #353 control regarding setting of balancing valve (V1).

3. FLOOR HEATING (2nd FLOOR): Upon a call for heat from 2nd floor master thermostat (T2), boiler firing is enabled under the control of the tekmar #250 boiler reset control (see suggested settings below) which is powered up through relay contact (R2-1). Circulator (P2) is also turned on through relay contact (R2-2). Boiler water is circulated through the 2nd floor distribution piping. Each heat emitter is equipped with its own thermostatic radiator valve (V2, V3, V4, V5, V6). As the temperature of a room rises slightly above its valve's setpoint, the valve slowly closes preventing further flow through that heat emitter. During occupied periods, the setting on thermostat (T2) should be slightly above the normal comfort temperature to maintain circulation in the 2nd floor distribution piping, allowing each room to be controlled by its thermostatic radiator valve. The 2nd floor master thermostat can be turned down to limit the temperature during unoccupied periods.

4. INITIAL CONTROL SETTINGS:
   a. 1st floor thermostat (T1) = 70 °F
   b. 2nd floor master thermostat (T2) = 75 °F
   c. Freeze prevention thermostat (TE) - 55 °F
   d. Boiler limit control (C1) = 170 °F
   e. tekmar #250 control (also see data sheet supplied with this control)
      Heating curve = 1.6
      Warm weather shutdown (WWSD) temperature = 70 °F
      Minimum supply temperature = 130 °F
      Differential = 15 °F
   f. Tekmar #353 control (also see data sheet supplied with this control)
      Heating curve = 0.5
      Occupied temperature = 70 °F
      Supply/return switch = return
      Maximum supply temperature = 120 °F

Figure 8-13 Sample System Operation Description

8•6 Piping Symbols

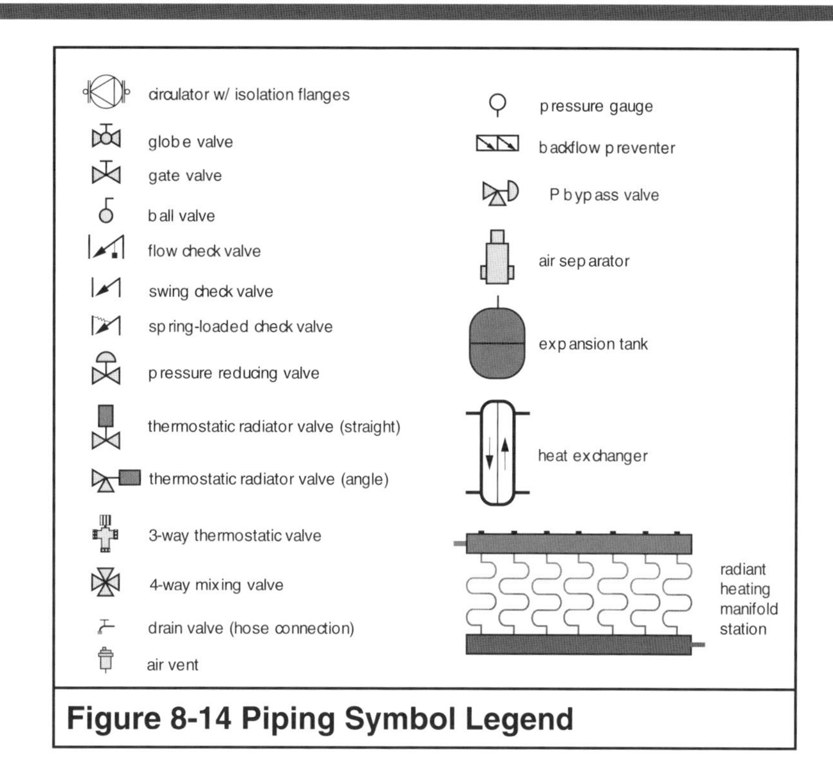

Figure 8-14 Piping Symbol Legend

Specifying and Sequencing

9•1 Introduction

Since there are many ways that a radiant heating system can be successfully installed, radiant designers or contractors may recommend or participate in the choice of which form of radiant heat to use. Many projects use different methods of installing radiant heat in different portions of a project. For example hydronic tubing might be put in a slab in a garage and basement but thin slab, plate, hanging or panel radiator or electric radiant systems might be installed elsewhere. This section will assist the specifying and sequencing decisions.

9•2 Specifying

The experience and expertise of a qualified contractor matters since the best system is often what the best available contractor feels most comfortable installing. A system correctly installed by a reliable and experienced contractor may be better than a theoretically more perfect system installed by a less knowledgeable or less experienced contractor.

The initial decision to use a hydronic or electric system is usually made on the basis of energy costs, ease of control and often on the scale of the job. Hydronic heating can utilize any source of energy that will make warm water. Water temperature can be controlled by varying temperature, flow, duration of flow and any combination of the three. Hydronic systems can also integrate domestic water heating, spa, pools, baseboard and radiators. Hydronic systems could change fuel sources in the future while electric systems are limited to electricity. Electric systems are very useful for spot heating small areas, for back up floor warming and for large projects in areas with assured low electrical costs.

If a decision to use a hydronic system is made there are other decisions that need to be made as to the suitability of the different system choices for different areas of a project. Costs of each type of system are not the same and need to be considered as well as acceleration, thermal mass, system R-value, structure heat loss, use patterns, and the preferences of the occupants.

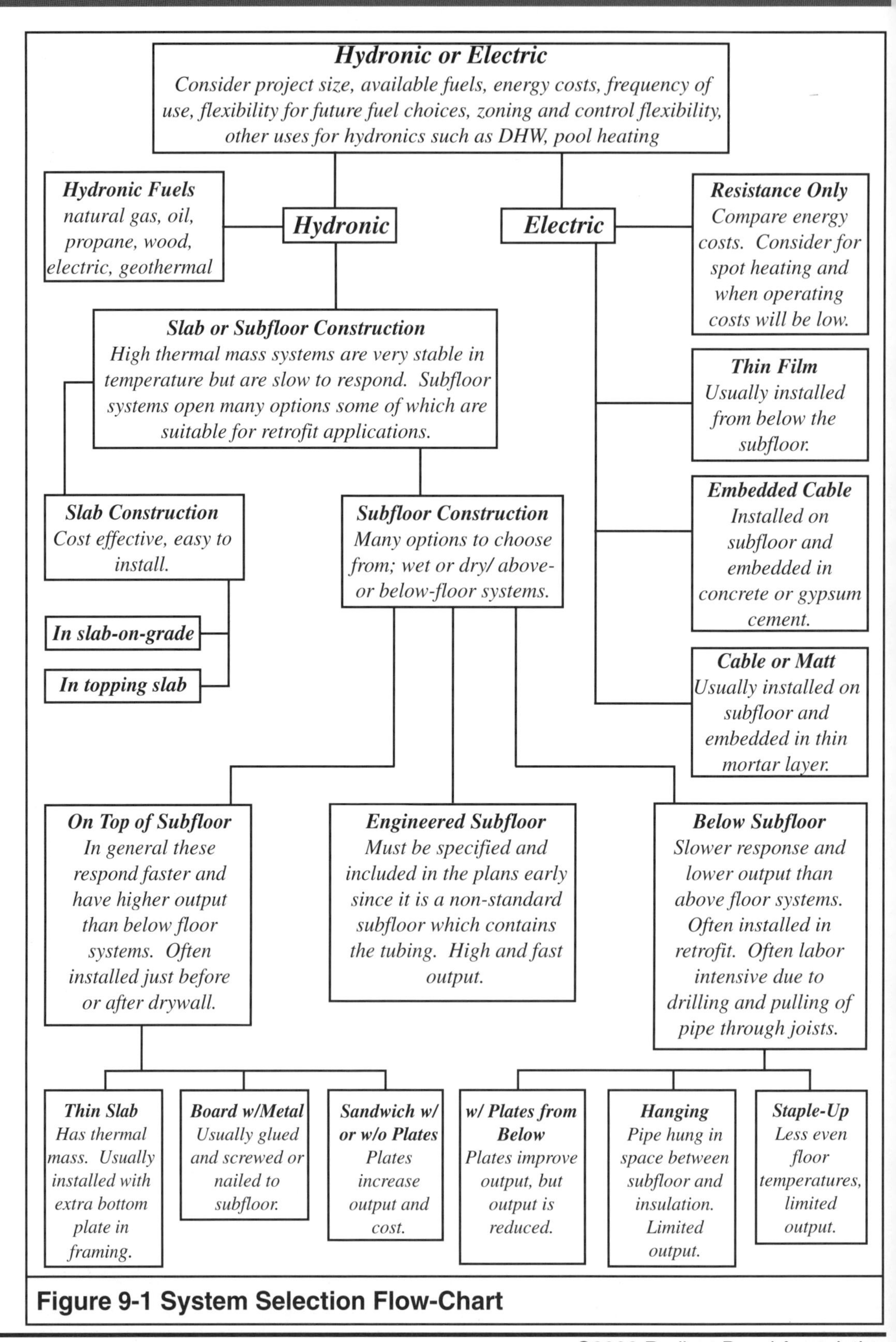

Figure 9-1 System Selection Flow-Chart

9

Acceleration

Acceleration measures how fast a system will accomplish a change in temperature. Radiant systems that accelerate rapidly are not better or worse than ones that accelerate slowly, but they do allow for the use of programmed set backs without a long lag time until the system responds. Systems that accelerate rapidly generally have a lower thermal mass than those that accelerate more slowly. Sometimes the acceleration is slowed by the placement of the system. For example, a system with hanging pipe in a joist space will respond more slowly than one with pipe placed above the subfloor because it must overcome the resistance of additional layers of material such as the subfloor.

SYSTEM TYPE	RATING
Slab On Grade (hydronic)	1
Thin Slab (hydronic)	3
Hanging Or Staple Up (hydronic)	1
Plates From Below Subfloor (hydronic)	3
Plates From Above Subfloor (hydronic)	4
Sandwich Above Subfloor (hydronic)	3
Structural Thermal Subfloor W/Metal (hydronic)	5
Boards With Metal Above Subfloor (hydronic)	4
Panel Radiator (hydronic)	5
Radiant Ceiling (hydronic)	4
Radiant Wall (hydronic)	4
Embedded Cable, Film or Matt Under Tile (electric)	5
Film Under Subfloor (electric)	1
KEY: 1=Slowest 5=Fastest	

Figure 9-2 Rate of Acceleration

Thermal Mass

Thermal mass is a measure of how many BTU's are contained within the system. High thermal mass systems tend to be very stable in temperature but are difficult to move up and down in temperature rapidly. In many instances this is a benefit. Thermal mass tempers solar gains, and can result in a stable temperature environment. There are a few circumstances when it is not a benefit. For example, a structure where there is likely to be a significant cooling load and a significant heating load on the same day for significant portions of the year. In this case a high mass system takes both energy to heat up and then cool down due to the time lags. High thermal mass systems are less suitable for set back temperature controls due to the time lags. They are very popular whenever cement is being poured as part of normal construction because of the low cost and ease of installation.

SYSTEM TYPE	RATING
Slab On Grade (hydronic)	5
Thin Slab (hydronic)	4
Hanging Or Staple Up (hydronic)	2
Plates From Below Subfloor (hydronic)	3
Plates From Above Subfloor (hydronic)	2
Sandwich Above Subfloor (hydronic)	2
Structural Thermal Subfloor W/Metal (hydronic)	3
Boards With Metal Above Subfloor (hydronic)	2
Panel Radiator (hydronic)	1
Radiant Ceiling (hydronic)	4
Radiant Wall (hydronic)	4
Embedded Cable, Film or Matt Under Tile (electric)	2
Film Under Subfloor (electric)	2
KEY: 1=Least 5=Most	

Figure 9-3 Quantity of Thermal Mass

Assembly R-value

The assembly R-Value measures the resistance to heat transfer of a delivery system, not including the R-Value of the coverings. These must be added together to get the total system R-Value. Assemblies with a high R-value will be limited by the additional R-value of the coverings and

9

limited in potential system output. For example, a hanging type system might not be able to deliver the required heat, particularly with thick carpet in a high heat loss area.

SYSTEM TYPE	RATING
Slab On Grade (hydronic)	2
Thin Slab (hydronic)	2
Hanging Or Staple Up (hydronic)	5
Plates From Below Subfloor (hydronic)	4
Plates From Above Subfloor (hydronic)	3
Sandwich Above Subfloor (hydronic)	4
Structural Thermal Subfloor W/Metal (hydronic)	1
Boards With Metal Above Subfloor (hydronic)	2
Panel Radiator (hydronic)	1
Radiant Ceiling (hydronic)	2
Radiant Wall (hydronic)	2
Embedded Cable, Film or Matt Under Tile (electric)	1
Film Under Subfloor (electric)	5
KEY: 1=Lowest 5=Highest	

Figure 9-4 Assembly Resistance

Retrofit Installations

Choosing the correct hydronic retrofit system for a remodeling project involves either applying the system below a sub floor, if there is access from below, or adding to the thickness of the floor from above to accommodate the radiant heating system. Systems applied from below limit the output of the radiant heating system due to the R-value of the subfloor layer. The R-value of the floor coverings placed on top further limits the system's heat output.

In any event an accurate heat loss and hydronic design must be done to verify the feasibility of the system. Many older homes are poorly insulated and pose more of a design challenge. The chart below gives additional information on the different systems

Zoning considerations

Hydronic systems are easy and relatively inexpensive to zone. This is normally accomplished by a thermostat that indirectly controls the flow of warm water

SYSTEM TYPE	
Slab On Grade (hydronic)	difficult due to thickness of slab and removal of floor coverings
Thin Slab (hydronic)	requires removal of floor coverings, adds 1.25"-1.5" to floor
Hanging Or Staple Up (hydronic)	requires access from below and floor assembly of modest R-value
Plates From Below Subfloor (hydronic)	requires access from below and floor assembly of modest R-value
Plates From Above Subfloor (hydronic)	requires removal of floor coverings, adds .75" to floor to floor thickness, reinstalling floor coverings
Sandwich Above Subfloor (hydronic)	requires removal of floor coverings, adds .75" to floor to floor thickness, reinstalling floor coverings
Structural Thermal Subfloor W/Metal (hydronic)	difficult unless flooring removed and installed on top of subfloor
Boards With Metal Above Subfloor (hydronic)	requires removal of floor coverings, adds .5"-.75" to floor thickness, reinstalling floor coverings
Panel Radiator (hydronic)	supply pipes must be routed in walls or under floors
Radiant Ceiling (hydronic)	requires removing sheetrock or soffiting and new sheetrock
Radiant Wall (hydronic)	requires removing sheetrock, or furring out and new sheetrock
Embedded Cable, Film or Matt Under Tile (electric)	requires removing floor coverings and embedding in grout on subfloor, reinstalling floor coverings
Film Under Subfloor (electric)	requires access from below and floor assembly of modest R-value
	notes are typical but may not apply to your project

Figure 9-5 Retrofit Considerations

to an area of a project. Zone areas together that have similar use, similar heat losses and similar temperature settings. Avoid zoning areas together that have dissimilar use patterns, heat losses or temperature settings. Zoning a hydronic system is different than balancing the loops of tubing within a zone. In a zone that has multiple runs or loops of tubing, adjustments can be made to the flow of water in each loop that will effect the heat output of the loop relative to other parts of the zone. Therefore, it is possible to make a hallway cooler than a dining room within a zone if the runs of tubing are configured and flows adjusted to accomplish this. Many Hydronic systems are over zoned when the same comfort could be achieved by balancing and adjusting flows.

Zone By Similar Heat Loss
Zone By Similar Use
Zone By Similar Set Points And Setback Times
Zone By Similar Environmental Orientation
Zone By Separate Floors
With Hydronics, Reduce Zoning By Adjusting Flows
THERMOSTAT CONSIDERATIONS
Place Thermostats In Protected Locations
Setbacks Are less Useful With High Thermal Mass
Floor Warming Systems May Use Floor Sensor Instead

Figure 9-6 Zoning

vey has not been done on the industry. The results in Figure 9-7 are based on a survey released in September of 1999 by the Radiant Panel Association. Percentages of market share do give strong clues as to cost and ease of installation. They do not indicate trends. Recent trends indicate a decline in the use of plates and growth in the use of board type systems. The popularity of slab on grade is due to the low additional cost of installing tubing when cement is already being poured. Thin slabs often add fireproofing and sound proofing advantages to a system. Many of the other systems have specific uses that should not be overlooked simply because they are not the most popular.

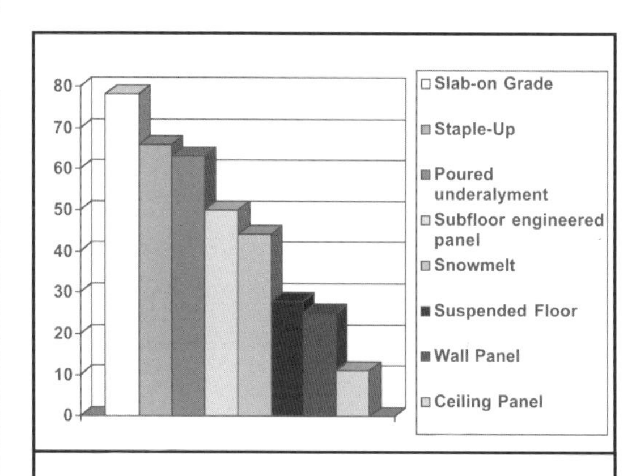

Figure 9-7 Percent of Installed Systems (RHR 2002, P&M July, 2002)

Market Share

Since the relative costs of most of the radiant systems differ and are very project specific, it is difficult to compare their costs. However, the pattern of how the market has selected systems in the past does give some indications as to cost and ease of installation.

A scientific, statistically accurate sur-

Electric Radiant Systems

Floor warming with electric radiant is an important and growing application. In this use the radiant system is not intended to heat the home but merely to provide comfortable warm floors. Typical uses are in tiled bathrooms or kitchens.

Many Electric radiant systems are installed in a grout layer between the sub floor

9

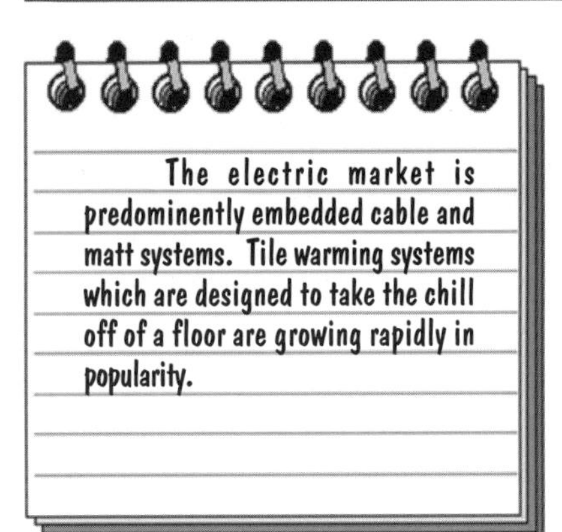

and tile. This allows the floor warming system to accelerate rapidly and be used on a programmable timer to provide warm floor comfort at times when it will be needed such as a chilly morning. Often they are installed with floor sensors that will limit and maintain a specific floor temperature. Since there is no intention to heat the space, the floor temperature only needs to be warm enough to add comfort. These systems have a wide application even in the southern climates where tile is used extensively but heating loads in the winter are modest.

Electric radiant can be used for spot radiant heating small areas in a project that uses another heating source for primary heating. Electric radiant can also be an effective choice for the primary heating system in a project. As with any heat source, the cost effectiveness and versatility of an electric system should to be compared with other systems. Utility costs, both present and future should be taken into consideration. Radiant slabs with thermal mass can take advantage of off peak discounts for electricity that coincide with the project use patterns. Electric systems do not require a large mechanical space or special accommodations for pipe or ducting, but the electrical service may need to be upgraded to meet the additional service load. They generally respond faster than hydronic systems and can accommodate almost any zoning configuration.

9•3 Sequencing

Proper sequencing of the installation allows for a quality radiant system with minimal installation problems. Coordination among the various trades is essential for a trouble-free radiant heating system installation.

Coordinate the following often overlooked items with the appropriate trades on the project:

- any under slab piping
- insulation requirements
- any pass-throughs required in the stem walls for supply and return pipes
- placement and installation of supply and return piping
- sequence of pipe, radiator or electric cable installation
- size of required electric service entrance and panel
- placement and installation of system and control wiring
- mechanical room requirements for combustion air, power, flue, floor drain, water supply, clearances, back flow prevention.
- routing of heating pipes to minimize conflicts with other trades
- any structural or loading issues associated with particular system types such as the weight of thin slabs or the added thickness of a board system on top of a sub floor
- verify that stairway rise/run calculations, floor drains and toiler flange installations take this into account any elevation in finished-floor height (frequently thin slab applications are framed with an extra bottom plate)

9

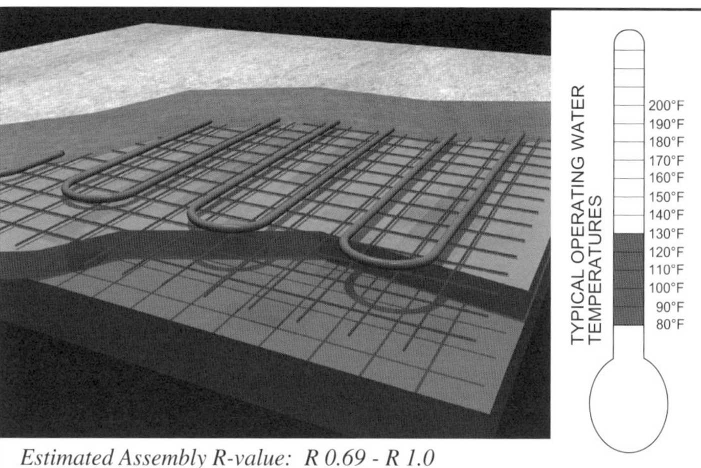

Estimated Assembly R-value: R 0.69 - R 1.0

Description

Radiant tubing is embedded in cement. The tubing is typically attached to metal mesh with plastic ties. A 4" slab is most typical. The tubing is best placed in the middle of the slab. Full under slab insulation is recommended for most residential applications. Slabs have a large thermal mass which stabilizes temperature swings but slows response. This method recommended whenever a slab is poured.

Very cost effective when cement is being poured.

Installation Notes: Most manufacturers recommend attaching tubing to steel mesh every 2 to 3 feet on straight runs and more frequently on corners. Pipe should be sleeved across expansion joints and where exiting cement.

Structural, Dimensional, and Material Requirements: Use closed cell insulation suitable for use under slabs. Embedding tubing in 4" of cement does not significantly weaken slab but unusual point loads should be referred to a structural engineer. Using round stone such as pea gravel in cement lesson chances of damage to tubing. Insulate under slabs to RPA guidelines and local codes.

System Output and Limitations: Unless limited by resistant floor coverings up, to 42 Btuh/sq.ft. in main use areas and up to 46 Btuh/sq.ft. in non traffic border areas.

Stage 1 - Planning and Design: Accurate heat loss and design with consideration of both delivery assembly R-values and specific floor covering R-values. Locate manifolds, thermostats and determine mechanical room requirements.

Stage 2 - Ground Work: Correctly size and run supply and return lines to maniifolds if under slab. Run mechanical room drain if required.

Stage 3 - Prior to Cement Pour: Install vapor barrier and insulation as required. Install tubing from manifold locations. Check spacing and maximum lengths. Pressure test and inspections. Maintain test during pour and construction.

Stage 4 - Rough Framing: Frame for manifold openings, mechanical room air supply and flue.

Stage 5 - Prior to Cover: Install supply and return lines to manifolds if in framing. Install Thermostat, zone valve and system electrical requirements. Install mechanical room water supply with backflow prevention, drain, air supply and flue venting. Inspections as required.

Stage 6 - Trim Out: Install manifolds if not previously installed. Install thermostats, zone valves, pumps, heating appliance and controls. Plumb mechanical room. Inspections as required.

Stage 7 - Start-up: Flush system, purge air from lines, add antifreeze if necessary, check correct sequencing and operation of all controls. Balance flow of system.

Stage 8 - Floor Covering: Run system to drive out all moisture from slab before installation of floor coverings.

Figure 9-8 Hydronic Slab-on-Grade

TYPICAL OPERATING WATER TEMPERATURES

200°F
190°F
180°F
170°F
160°F
150°F
140°F
130°F
120°F
110°F
100°F
90°F
80°F

Estimated Assembly R-value: R 0.69 - R 1.0

Description

Radiant tubing is attached to top of slab with plastic clips, track system or wire mesh. A thin slab of cement is poured over the tubing. Typical slabs are 1.5" thick when using 1/2" tubing but other thicknesses are possible. This system is useful over engineered slabs such as seismic or post tensioned slabs. Refer these to an engineer. Often used in retrofit applications.

Cost varies widely due to size and job location.

Installation Notes: Most manufacturers recommend attaching tubing every 18" plus more on the corners. Pipe should be rigid, non-compressable insulation is often used between slab and topping slab. Clips are available for attaching tube to rigid insulation.

Structural, Dimensional, and Material Requirements: Topping slabs should only be installed on a stable underlying slab that is in good condition. Rigid insulation should be of a type and compressive strength suitable for use under a slab. Tubing should be sleeved where it passes through expansion joints. Coverage of tubing with cement is usually at least 3/4" over the top of the tubing.

System Output and Limitations: Unless limited by resistant floor coverings up to 42 Btuh/sq.ft. in main use areas and up to 46 Btuh/sq.ft. in non-traffic border areas.

Stage 1 - Planning and Design: Accurate heat loss and design with consideration of both delivery assembly R-values and specific floor covering R-values. Locate manifolds, thermostats and determine mechanical room requirements. Refer any structural slab issues to an engineer.

Stage 2 - Ground Work: Correctly size and run supply and return lines to manifolds if under slab. Run mechanical room drain if required.

Stage 3 - Prior to Tubing Installation: Do not use petroleum products in contact with pipe such as tar paper. Check insulation type if used under topping slab. Check tube spacing and maximum lengths. Pressure test and inspections. Maintain test during pour and consturction.

Stage 4 - Rough Framing: Frame for manifold openings, mechanical room air supply and flue.

Stage 5 - Prior to Cover: Install supply and return lines to manifolds if in framing. Install Thermostat, zone valve and system electrical requirements. Install mechanical room water supply with backflow prevention, drain, air supply and flue venting. Inspections as required.

Stage 6 - Trim Out: Install manifolds if not previously installed. Install thermostats, zone valves, pumps, heating appliance and controls. Plumb mechanical room. Inspections as required.

Stage 7 - Start-up: Flush system, purge air from lines, add antifreeze if necessary, check correct sequencing and operation of all controls. Balance flow of system.

Stage 8 - Floor Covering: Run system to drive out all moisture from slab before installation of floor coverings

Figure 9-9 Hydronic Topping Slab

9

Estimated Assembly R-value: R 0.69 - R 1.0

Description

Radiant tubing is attached to top of subfloor with approved staples or plastic clips. A thin slab of gypsum based cement or cement is poured over the tubing. Typical slabs are 1.5" thick when using 1/2" tubing but may be as thin as 1.25" thick when using 3/8" tubing. Gypsum cement is lighter than cement, but a little less conductive. It bonds to the subfloor while cement is separted with a slip sheet. Cost varies widely due to size and job location.

Installation Notes: Most manufacturers recommend attaching tubing with clips or staples every 18" plus more on the corners. Pipe should be sleeved across expansion joints and where exiting cement. Insulate to RPA guidelines between floors and insulate to RPA guidelines and meet code over unheated space. Usually requires pump truck.

Structural, Dimensional, and Material Requirements; Joists must be designed to carry the loads of the cement or gypsum cement. Typically cement at 1.5" thickness weighs 17.5 lbs. per sq/ft, gypsum cement 13.5 lbs. per sq/ft. Pumping 6 sack mix with pea gravel as stiff as possible has been successfull, with or without addition of fiber. Typically, walls are framed with double bottom plate. Fill holes in floor, install sleepers at floor transitions and for carpet tack strips.

System Output and Limitations; Unless limited by resistant floor coverings up to 42 Btuh/sq.ft. in main use areas and up to 46 Btuh/sq.ft. in non traffic border areas.

Stage 1 - Planning and Design: Accurate heat loss and design with consideration of both delivery assembly R-values and specific floor covering R-values. Locate manifolds, thermostats and determine mechanical room requirements.

Stage 2 - Ground Work: Correctly size and run supply and return lines to manifolds if under slabs, stem walls, etc. Run mechanical room drain if required.

Stage 3 - Prior to Cement Pour: Sealing subfloor with PVA is frequently specified. Do not use petroleum products in contact with pipe such as tar paper. Check tube spacing and maximum lengths. Pressure test and inspections. Maintain test during pour and construction.

Stage 4 - Rough Framing: Frame for manifold openings, mechanical room air supply and flue.

Stage 5 - Prior to Cover: Install supply and return lines to manifolds if in framing. Install Thermostat, zone valve and system electrical requirements. Install mechanical room water supply with backflow prevention, drain, air supply and flue venting. Inspections as required.

Stage 6 - Trim Out: Install manifolds if not previously installed. Install thermostats, zone valves, pumps, heating appliance and controls. Plumb mechanical room. Inspections as required.

Stage 7 - Start-up: Flush system, purge air from lines, add antifreeze if necessary, check correct sequencing and operation of all controls. Balance flow of system.

Stage 8 - Floor Covering: Run system to drive out all moisture from slab before installation of floor coverings.

Figure 9-10 Thin-Slab on Subfloor

Estimated Assembly R-value: R 1.7 - R 2.2 (pipe w/ 3/4" plywood)

Description

Radiant tubing is hung or attached to the underside of the joists in an airspace with insulation below. Requires higher water temperatures and has more limited heat output than other systems. It is often used for retrofitting when access from below is possible. Hanging systems have more even joist cavity temperatures than when pipe is attached in contact with the subfloor or joists.

Low material costs; labor intensive installation.

Installation Notes: Most manufacturers recommend hanging or attaching tubing every 18"; more closely on corners. Tubing should be secured at corners and where penetrating joists. Note: Some jurisdictions do not allow plenum heating systems due to fire code flame spread issues. Check with authorities before specifying or installing.

Structural, Dimensional, and Material Requirements: Holes (typically 1-3/8") need to be drilled in joists for tubing . Insulate to RPA guidelines between floors and to meet code over unheated space. Controlled research has not been done on any problems associated with elevated or uneven joist temperatures, but there are many successful systems in operation.

System Output and Limitations: More limited than other systems in output (up to 28 Btuh/sq. ft. or significantly less depending on floor coverings). Limit floor covering R-value or reduce structure heat loss or both to insure system effectiveness.

Stage 1 - Planning and Design: Accurate heat loss and design with particular consideration of delivery assembly R-values and specific floor covering R-values. Locate manifolds, thermostats and determine mechanical room requirements.

Stage 2 - Ground Work: Make pass-throughs in stem walls if required for supply and return lines. Correctly size and run supply and return lines to manifolds if under slab or stem walls. Provide for future mechanical room drain.

Stage 3 - Prior to Tubing Installation: Locate manifolds. Layout runs of tubing and develop drilling plan for joists. Drill holes in joists as required. Attach tubing securely.

Stage 4 - Rough Framing: Install tubing in joist space. Check loop lengths and spacing. Check for secure attachment of tubing. Pressure test and maintain tubing under test during construction. Frame for manifold openings, mechanical room air supply and flue.

Stage 5 - Prior to Cover: Install supply and return lines to manifolds if in framing. Maintain pressure test on system parts that will be under cover. Install thermostat, zone valve and system electrical requirements. Install mechanical room water supply with backflow prevention, drain, air supply and flue venting. Inspections as required.

Stage 6 - Trim Out: Install manifolds if not previously installed. Install thermostats, zone valves, pumps, heating appliance and controls. Plumb mechanical room. Inspections as required.

Stage 7 - Start-up: Flush system, purge air from lines, add antifreeze if necessary, check correct sequencing and operation of all controls. Balance flow of system.

Stage 8 - Floor Covering: As the output of hanging systems is limited, use as conductive a floor covering as possible.

Figure 9-11 Hanging or Attached Below Subfloor

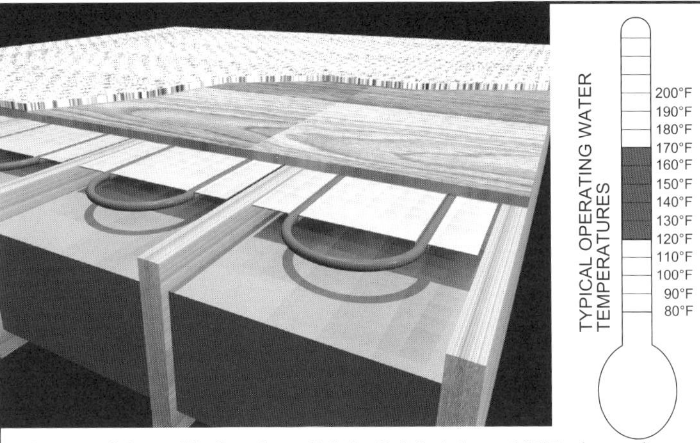

Description

Radiant tubing is attached to the underside of the joists with metal plates to diffuse the heat. Insulation is recommended below the plates. Higher water temperatures and more limited heat output than above subfloor systems, but plates make it more effective than hanging pipe from under joists. It is often used for retrofitting when access to joist space is available.

Low material cost; labor intensive installation

Estimated Assembly R-value: R 1.3 - R 1.8 (pipe w/ 3/4" plywood)

Installation Notes: Noise from expansion of pipe moving in plates has been occasional problem, particularly with barrier pipe. This can be prevented by installing a thin sheet of polyethylene plastic between pipe and plates. Plates can be labor intensive to install Tubing should be well secured at corners and where penetrating joists.

Structural, Dimensional, and Material Requirements: Holes (typically 1-3/8") need to be drilled in joists for tubing. Insulate to RPA guidelines between floors and to meet code over unheated space. Plates need to be secured to underside of subfloor.

System Output and Limitations: Unless limited by resistant floor coverings, up to 42 Btuh/ sq. ft. in main use areas, and up to 46 Btuh/sq. ft. in non-traffic border areas.

Stage 1 - Planning and Design: Accurate heat loss and design with consideration of both delivery assembly R-values and specific floor covering R-values. Locate manifolds, thermostats and determine mechanical room requirements.

Stage 2 - Ground Work: Make pass-throughs in stem walls if required for supply and return lines. Correctly size and run supply and return lines to maniifolds if under slab or stem walls. Provide for future mechanical room drain.

Stage 3 - Prior to Cement Pour: Locate manifolds. Layout runs of tubing and develop drilling plan for joists. Drill holes in joists as required. Attach tubing securely

Stage 4 - Rough Framing: Install tubing and plates in joist space. Check loop lengths and spacing. Check for secure attachment of tubing. and plates. Pressure test and maintain tubing under test during construction. Frame for manifold openings, mechanical room air supply and flue.

Stage 5 - Prior to Tubing Installation: Install supply and return lines to manifolds if in framing. Install Thermostat, zone valve and system electrical requirements. Install mechanical room water supply with backflow prevention, drain, air supply and flue venting. Inspections as required.

Stage 6 - Trim Out: Install manifolds if not previously installed. Install thermostats, zone valves, pumps, heating appliance and controls. Plumb mechanical room. Inspections as required.

Stage 7 - Start-up: Flush system, purge air from lines, add antifreeze if necessary, check correct sequencing and operation of all controls. Balance flow of system.

Stage 8 - Floor Covering: As the output of below subfloor systems is limited, use as conductive a floor covering as possible.

Figure 9-12 With Plates Below Subfloor

Estimated Assembly R-value: R 0.6

Description

Premanufactured panels have grooves for tubing and an aluminum sheet bonded to the board. In this case the premanu-factured panels serve both as the structural subfloor and as the channel into which the tubing is installed. The aluminum sheet makes the system accelerate rapidly and spreads out the heat. Tubing is normally installed 12" on center in grooves.

Cost is reduced since tubing is installed 12-inches on center and panel is the subfloor.

Installation Notes: This product often requires that tubing be installed early in a project. When this occurs the tubing must be protected from damage during the remainder of construction. The boards must be carefully aligned and laid out. A CAD layout of the placement panels is recommended.

Structural, Dimensional, and Material Requirements: Panels come in 4'x8' and can be cut with a saw. Nonstandard thickness of the subfloor must be accounted for in building elevations and construction. Some cut panels need blocking for support as recommended by manufacturer. Since this board is the subfloor, it must be installed throughout the entire radiant area.

System Output and Limitations: Unless limited by resistant floor coverings, up to 42 Btuh/sq. ft.in main use areas and up to 46 Btuh/sq. ft. in non-traffic border areas. This system operates with low water temperatures and with a high output.

Stage 1 - Planning and Design: Accurate heat loss and design with consideration of both delivery assembly R-values and specific floor covering R-values. Panel CAD layout recommended. Locate manifolds, thermostats and determine mechanical room requirements.

Stage 2 - Ground Work: Make pass-throughs in stem walls, if required, for supply and return lines. Correctly size and run supply and return lines to maniifolds if under slab or stem walls. Provide for future mechanical room drain.

Stage 3 - Prior to Tubing Installation: Locate manifolds. Layout runs of tubing and develop drilling plan for joists. Route any special pathways in panels. Drill holes in joists as required.

Stage 4 - Rough Framing: Normally install tubing in panels with caulking. Check loop lengths and spacing. Check for secure attachment of tubing. Pressure test and maintain tubing under test during construction. Protect pipe from physical and UV damage Frame for manifold openings, mechanical room air supply and flue.

Stage 5 - Prior to Cover: Install supply and return lines to manifolds if in framing. Install thermostat, zone valve and system electrical requirements. Install mechanical room water supply with backflow prevention, drain, air supply and flue venting. Inspections as required.

Stage 6 - Trim Out: Install manifolds if not previously installed. Install thermostats, zone valves, pumps, heating appliance and controls. Plumb mechanical room. Inspections as required.

Stage 7 - Start-up: Flush system, purge air from lines, add antifreeze if necessary, check correct sequencing and operation of all controls. Balance flow of system.

Stage 8 - Floor Covering: Some floor coverings require a layer of backerboard. Others may be installed directly over the tubing/subfloor combination per manufacturers recommendations.

Figure 9-13 Engineered Subfloor with Metal and Tubing Grooves

9

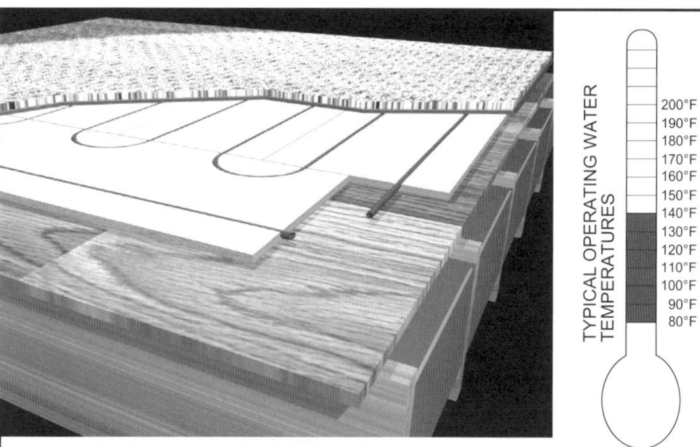

Estimated Assembly R-value: R 0.75 - R 1.1 (depending on product)

Description

Two varieties exist. One board has metal on the bottom and the other on the top. Both serve to spread the heat laterally. Normally they are glued and screwed or stapled to the top of a wooden subfloor. Under some conditions they may be attached on top of existing slabs. Modular system with straight and curved end pieces are assembled to make a channel for pipe. Different products use different pipe sizes. Cost may be reduced when board is only applied to heated areas.

Installation Notes: The boards must be carefully aligned and laid out. Spacing of pipe, board size and thickness varies from product to product. Typically boards are 1/2" or 5/8" thick, making them useful in retrofit applications as well as new construction. They can be installed quite late in job sequence or just after sheetrock to provide job heat.

Structural, Dimensional, and Material Requirements: Thickness of boards needs to be accounted for in cabinet toe kicks, stair rise and runs. Some floor coverings may require a layer of backer board on top of radiant board system.

System Output and Limitations: Unless limited by resistant floor coverings, up to 42 Btuh/sq. ft. in main use areas and up to 46 Btuh/sq. ft. in non-traffic border areas. This system operates with low water temperatures and with a high output.

Stage 1 - Planning and Design: Accurate heat loss/design with consideration of both delivery assembly R-values and specific floor covering R-values. CAD layout of boards is helpful, not required. Locate manifolds, thermostats and determine mechanical room requirements.

Stage 2 - Ground Work: Make pass-throughs in stem walls if required for supply and return lines. Correctly size and run supply and return lines to manifolds if under slab or stem walls. Provide for future mechanical room drain.

Stage 3 - Prior to Tubing Installation: Locate manifolds. Layout runs of tubing and develop drilling plan for joists. Route any special pathways in panels. Drill holes in joists as required.

Stage 4 - Rough Framing: Normally install tubing in panels with approved silicone caulking. Check loop lengths and spacing. Check for secure attachment of tubing. Pressure test and maintain tubing under test during construction. Protect pipe from physical and UV damage Frame for manifold openings, mechanical room air supply and flue.

Stage 5 - Prior to Cover: Install supply and return lines to manifolds if in framing. Install thermostat, zone valve and system electrical requirements. Install mechanical room water supply with backflow prevention, drain, air supply and flue venting. Inspections as required.

Stage 6 - Trim Out: Install manifolds if not previously installed. Install thermostats, zone valves, pumps, heating appliance and controls. Plumb mechanical room. Inspections as required.

Stage 7 - Start-up: Flush system, purge air from lines, add antifreeze if necessary, check correct sequencing and operation of all controls. Balance flow of system.

Stage 8 - Floor Covering: Some floor coverings require a layer of backerboard. Others may be installed directly over the tubing/board combination per manufacturers recommendations.

Figure 9-14 Above-Floor Boards with Metal and Grooves

Description

Typically 1" x 4" x 3/4" nailing sleepers are attached to the top of the subfloor and pipe is placed in between the sleepers with or without the addition of the metal plates. The metal plates which typically cover about 80% of the pipe, add significantly to the evenness of heat, and provide a higher output and faster acceleration.

Cost of plates can add significantly to the system price.

Estimated Assembly R-value: R 0.75 - R 1.15 (depending if plates are used)

Installation Notes: Noise from expansion of pipe moving in plates has been an occasional problem, particularly with barrier pipe. This can be prevented by installing a thin sheet of polyethylene plastic between pipe and plates. Plates can be labor intensive to install. Tubing should be well secured at corners.

Structural, Dimensional, and Material Requirements: Thickness of boards needs to be accounted for in cabinet toe kicks, stair rise and runs. Some floor coverings may require an additional layer of plywood or backer board on top of radiant board systems.

System Output and Limitations: With plates: unless limited by resistant floor coverings, up to 42 Btuh/sq. ft. in main use areas and, up to 46 Btuh/sq. ft. in non traffic border areas. Without plates: output and acceleration more limited .

Stage 1 - Planning and Design: Accurate heat loss/design with consideration of both delivery assembly R-values and specific floor covering R-values. Locate manifolds, thermostats and determine mechanical room requirements.

Stage 2 - Ground Work: Make pass-throughs in stem walls if required for supply and return lines. Correctly size and run supply and return lines to maniifolds if under slab or stem walls. Provide for future mechanical room drain.

Stage 3 - Prior to Tubing Installation: Locate manifold locations. Layout runs of tubing and install sleepers. Cut any special pathways in furring Drill holes in framing as required.

Stage 4 - Rough Framing: Normally install tubing in sleepers. Check loop lengths and spacing. Check for secure attachment of tubing. Pressure test and maintain tubing under test during construction. Protect pipe from physical and UV damage. Frame for manifold openings, mechanical room air supply and flue.

Stage 5 - Prior to Cover: Install supply and return lines to manifolds if in framing. Install Thermostat, zone valve and system electrical requirements. Install mechanical room water supply with backflow prevention, drain, air supply and flue venting. Inspections as required.

Stage 6 - Trim Out: Install manifolds if not previously installed. Install thermostats, zone valves, pumps, heating appliance and controls. Plumb mechanical room. Inspections as required.

Stage 7 - Start-up: Flush system, purge air from lines, add antifreeze if necessary, check correct sequencing and operation of all controls. Balance flow of system.

Stage 8 - Floor Covering: Some floor coverings require a layer of plywood or backerboard. Others may be installed directly over the tubing/board combination per manufacturers recommendations.

Figure 9-15 Sandwich Method with or without Plates

9

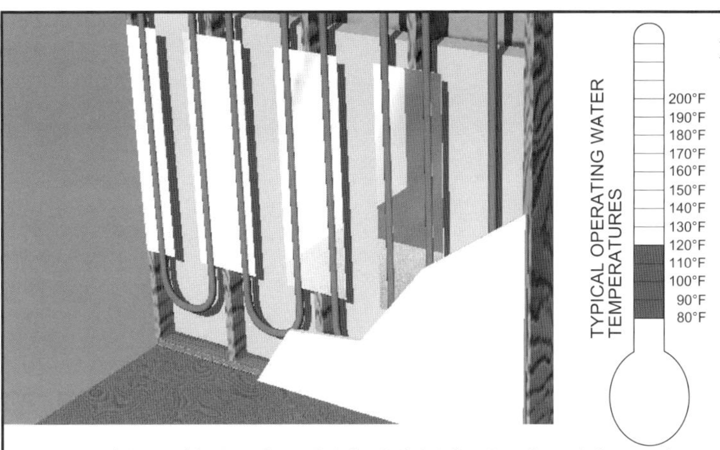

Estimated Assembly R-value: R 0.8 - R 1.3 (plus R-value of sheetrock)

Description

Typically metal plates are attached to framing or furring strips, holes are drilled through the framing, The pipe is threaded through the framing and snapped into the grooves of the plates. Plates typically cover about 80% of the pipe and add significantly to the heat and provide a higher output and faster acceleration. Often used to add additional heat in conjunction with radiant floors.

Cost of plates adds significantly to the system price.

Installation Notes: Noise from expansion of pipe moving in plates has been occasional problem particularly with barrier pipe. This can be prevented by installing thin sheet of polyethylene plastic between pipe and plates. Plates can be labor intensive to install. Tubing should be well secured at corners. Attach plates on one side only.

Structural, Dimensional, and Material Requirements: Insulation should be used in walls for control purposes to prevent heating adjacent rooms unintenmtionally. Insulation should be up against back of plates.

System Output and Limitations: With Plates: unless limited by resistant wallcoverings, up to 45-50 Btuh/sq. ft. Systems are problematic at the heights where picture frames and other items are often nailed into wall. Water temperatures should be limited to 120 °F to prevent damaging or discoloring sheet rock.

Stage 1 - Planning and Design: Accurate heat loss/design with consideration of both delivery assembly R-values and specific wall covering R-values. Locate manifolds, thermostats and determine mechanical room requirements.

Stage 2 - Ground Work: Make pass-throughs in stem walls if required for supply and return lines. Correctly size and run supply and return lines to manifolds if under slab or stem walls. Provide for future mechanical room drain.

Stage 3 - Prior to Tubing Installation: Locate manifolds. Layout runs of tubing and install furring or drill framing for tube. Cut any special pathways in furring. Drill holes in framing and joists as required.

Stage 4 - Rough Framing: Check loop lengths and spacing. Check for secure attachment of tubing. Pressure test and maintain tubing under test during construction. Protect pipe from physical and UV damage. Frame for manifold openings, mechanical room air supply and flue.

Stage 5 - Prior to Cover: Install supply and return lines to manifolds if in framing. Install Thermostat, zone valve and system electrical requirements. Install mechanical room water supply with backflow prevention, drain, air supply and flue venting. Inspections as required.

Stage 6 - Trim Out: Install manifolds if not previously installed. Install thermostats, zone valves, pumps, heating appliance and controls. Plumb mechanical room. Inspections as required.

Stage 7 - Start-up: Flush system, purge air from lines, add antifreeze if necessary, check correct sequencing and operation of all controls. Balance flow of system.

Stage 8 - Wall Covering: Wall coverings other than normal thicknesses of sheet rock should be carefully addressed at design stage.

Figure 9-16 Radiant Wall with Plates

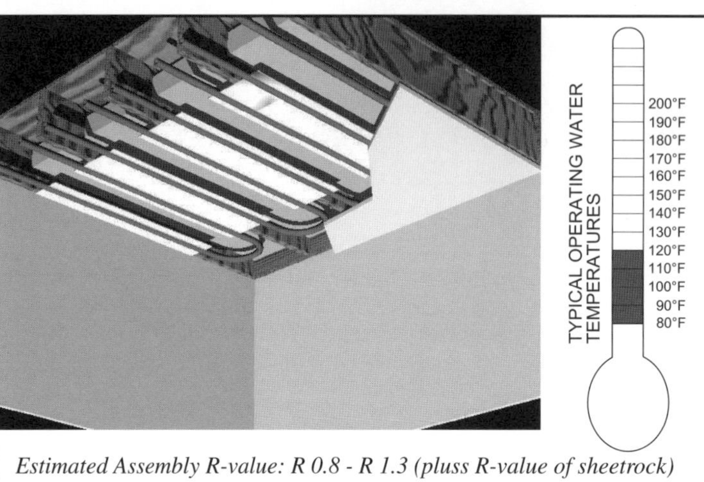

Description

Typically metal plates are attached to framing or furring strips. Holes are drilled through the framing. Pipe is threaded through the framing and snapped into the grooves of the plates. Plates typically cover about 80% of the pipe, add significantly to the heat and provide a higher output and faster acceleration. Can be used in retrofit by furring down or soffiting a ceiling. Cost of plates adds significantly to the system price.

Estimated Assembly R-value: R 0.8 - R 1.3 (pluss R-value of sheetrock)

Installation Notes: Noise from expansion of pipe moving in plates has been occasional problem, particularly with barrier pipe. This can be prevented by installing thin sheet of polyethylene plastic between pipe and plates. Plates can be labor intensive to install. Tubing should be well secured at corners. Attach plates on one side only.

Structural, Dimensional, and Material Requirements: Insulation should be used between floors for control purposes to prevent heating adjacent floors unintenmtionally. Insulation should be up against back of plates.

System Output and Limitations: With pates: unless limited by resistant ceiling coverings, up to 45-50 Btuh/ sq. ft. Water temperatures should be limited to 120 °F to prevent damaging or discoloring sheetrock.

Stage 1 - Planning and Design: Accurate heat loss/design with consideration of both delivery assembly R-values and specific ceilingcovering R-values. Locate manifolds, thermostats and determine mechanical room requirements.

Stage 2 - Ground Work: Make pass-throughs in stem walls if required for supply and return lines. Correctly size and run supply and return lines to manifolds if under slab or stem walls. Provide for future mechanical room drain.

Stage 3 - Prior to Tubing Installation: Locate manifolds. Layout runs of tubing and install furring or drill framing for tube. Cut any special pathways in furring. Drill holes in framing and joists as required.

Stage 4 - Rough Framing: Check loop lengths and spacing. Check for secure attachment of tubing. Pressure test and maintain tubing under test during construction. Protect pipe from physical and UV damage. Frame for manifold openings, mechanical room air supply and flue.

Stage 5 - Prior to Cover: Install supply and return lines to manifolds if in framing. Install thermostat, zone valve and system electrical requirements. Install mechanical room water supply with backflow prevention, drain, air supply and flue venting. Inspections as required.

Stage 6 - Trim Out: Install manifolds if not previously installed. Install thermostats, zone valves, pumps, heating appliance and controls. Plumb mechanical room. Inspections as required.

Stage 7 - Start-up: Flush system, purge air from lines, add antifreeze if necessary, check correct sequencing and operation of all controls. Balance flow of system.

Stage 8 - Ceiling Covering: Ceiling coverings other than normal thicknesses of sheet rock should be carefully addressed at design stage. Paint and texture is not affected.

Figure 9-17 Radiant Ceiling with Plates

9

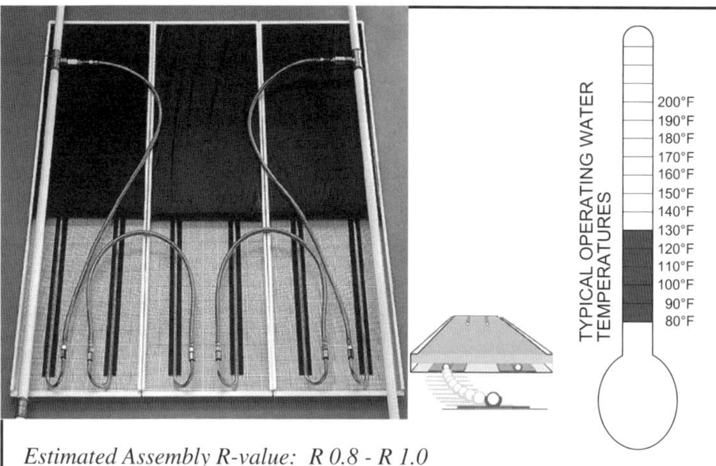

Estimated Assembly R-value: R 0.8 - R 1.0

Description

Typically modular hydronic metal ceiling plates are incorporated into a suspended commercial grade ceiling system. When used for cooling it must be accompanied with an air dehumidification system. The required air ducting will be much smaller than for a forced air heating and cooling system. The reduced size may even allow for extra floors in tall buildings of the same height.

Cost is more feasible for commercial projects.

Installation Notes: Normally units are placed in ceiling grid and connected to supply and return pipes with flexible pipe to facilitate service.

Structural, Dimensional, and Material Requirements: Ceiling grid must be strong enough to hold weight of units and water content.

System Output and Limitations: Units are normally sized by the cooling load which requires typically more panel area than heating. These are generally engineered systems where both the heating and cooling outputs per unit of area can be varied widely by changing the temperature difference between the water and the room temperature.

Stage 1 - Planning and Design: Accurate engineered heating and cooling design with consideration of cooling dehumidification needs. Locate piping, manifolds, thermostats and determine mechanical room equipment and requirements.

Stage 2 - Ground Work: Make pass-throughs in stem walls if required for supply and return lines. Correctly size and run supply and return lines to manifolds if under slab or stem walls. Provide for future mechanical room drain.

Stage 3 - Prior to Panel Installation: Locate panels. Install ceiling grid. Run supply and return piping.

Stage 4 - Rough Framing: Check panel layout and connections. Check for secure attachment of panel connectors. Pressure test and maintain tubing under test during construction. Protect pipe from physical and UV damage

Stage 5 - Prior to Cover: Install supply and return lines to panels if under cover. Install thermostat, zone valve and system electrical requirements. Install mechanical room water supply with backflow prevention, drain, air supply and flue venting. Inspections as required.and flue venting. Inspections as required.

Stage 6 - Trim Out: Install manifolds if not previously installed. Install thermostats, zone valves, pumps, heating appliance and controls. Plumb mechanical room. Inspections as required.

Stage 7 - Start-up: Flush system, purge air from lines, add antifreeze if necessary, check correct sequencing and operation of all controls. Balance flow of system.

Stage 8 - Floor Covering: Ceiling panels should not be covered or obstructed. Cooling requires a seperate dehumidification system and chilled water.

Figure 9-18 Modular Ceiling Radiant Heating and Cooling

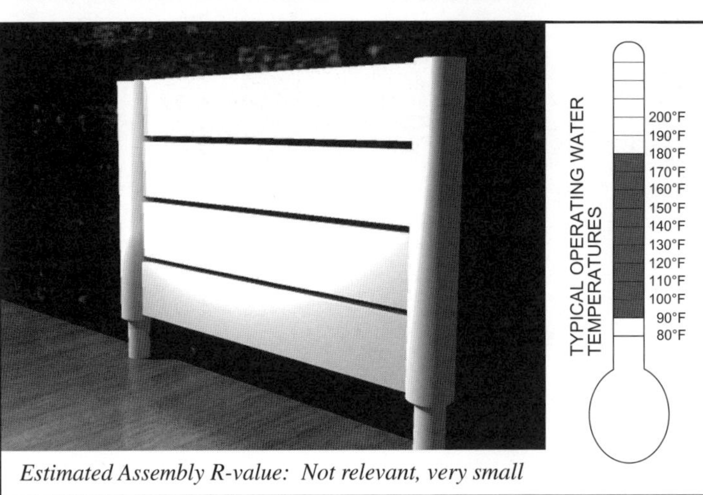

Estimated Assembly R-value: Not relevant, very small

Description

These modern radiators combine a significant radiant output with some convective output as warm air rises through them. They are available in many decorative shapes, colors and sizes. When properly sized they can operate at lower water temperatures than traditional hydronic baseboard.

Cost varies; decorative models are expensive.

Installation Notes: Typically require rough in blocking for attachment after sheetrock.

Structural, Dimensional, and Material Requirements: Will attach to most walls if blocking for support brackets are properly installed. Supply and return lines must be plumbed to each unit through framing in walls, floors, or sometimes, ceilings.

System Output and Limitations: Output dependent coverage area and operating water temperature. Can operate over wide temperature range.

Stage 1 - Planning and Design: Accurate heat loss and design showing size and placement of units. Routing for supply and return lines must be determined.
Stage 2 Ground Work: Run supply and return lines that will be under slabs or in ground.
Stage 3 - Piping Rough-in: Run all supply and return lines for radiators, pressure tests and inspections as requir
Stage 4 - Rough Framing: Install support blocking.
Stage 5 - Prior to Cover: Run all wiring for controls. Maintain pressure test on supply and return lines during phases of construction that might damage them.
Stage 6 - Trim Out: Install radiators, thermostats and controls.
Stage - 7 Start-up: Test all sequences of operation. Check for leaks.
Stage - 8 Floor Covering: Floor coverings are not an issue with these products.

Figure 9-19 Eurostyle Radiator

Description

The electric cable or matt is encapusulated by embedding it in a 1/8"-1/2" mud bed or in thin set mortar on top of the subfloor. When only floor-warming is desired, the system is controlled by a floor temperature sensor. Heating systems are controlled with a thermostat often in combination with a floor sensor.

Cost of installation is generally low.

Estimated Assembly R-value: Embeding layer R 0.2

Installation Notes: Product must be carefully and evenly embedded to protect the cable from point load damage. Cable systems must be evenly spaced to prevent hot spots in the cable.

Structural, Dimensional, and Material Requirements: Embedding layer usually 1/8"-1/2" mud bed or mortar must adhere to subfloor.

System Output and Limitations: Floor-warming systems are normally designed to produce 8-15 Btuh/sq. ft. Heating systems produce up to 45 Btuh/sq. ft. unless limited by the R-value of the floor coverings. Do not place under areas that will be covered by insulating objects such as zero clearance furniture or boxes of inventory unless approved by the manufacturer.

Stage 1 - Planning and Design: Accurate heat loss and design taking into consideration floor coverings. Make sure electric service panel is adequate. Locate sub panels, relay boxes and controls.

Stage 2 - Ground Work: Run any underground electric wiring and required service connections.

Stage 3 - Prior to Electric Radiant Installation: Make sure subfloor is clean, structurally sound and suitable as a substrata for embedding layer.

Stage 4 - During Electric Radiant Installation: If cable, make certain spacing is as recommended. If floor sensor is being used, make sure it is embedded in a representative location.

Stage 5 - Prior to Cover: Make sure all rough in wiring for electric radiant is installed. Perform continuity check on all circuits.

Stage 6 - Trim Out: Install thermostats and other controls. Protect embedding layer from abuse that could damage cable or matt.

Stage 7 - Start-up: Test all sequences of operation.

Stage 8 - Floor Covering: Use floor sensor under floor coverings such as vinyl that have temperature limitations. Some systems have limitation of R-value of coverings that may be placed on top. System accelerates rapidly when used with low R-value coverings.

Figure 9-20 Embedded Electric Cable and Mat

9

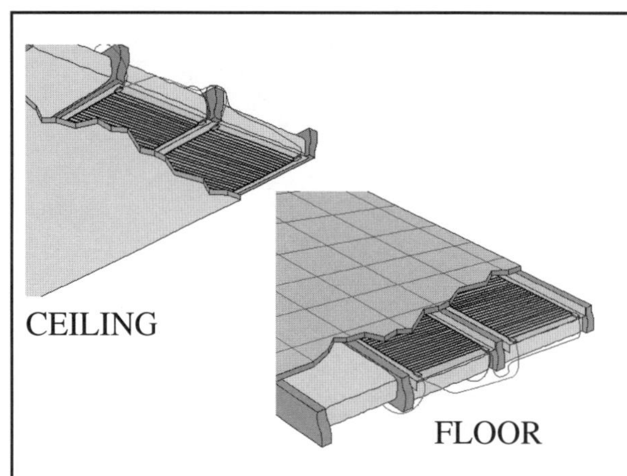

Description

Typically made of a polyester film where conductive ink elements are embedded within the film. The film is stapled between the joists. It is available in different widths and outputs. When used under floors, the film is staple 2" below sub floor and insulation placed beneath it. Lower output films are used for ceilings.

Cost of installation is generally low.

Estimated Assembly R-value: R 0.05 (ceilings), R 0.5 (floors with 2-inch air space)

Installation Notes: Do not staple element, staple only near areas of film. Use correct film width for joist spacing. Do not fold. Use approved connection system.

Structural, Dimensional, and Material Requirements: Will fit in most framing but requires standard joist widths.

System Output and Limitations: A variety of output of films are available. Ceiling panels may be installed without an airspace behind sheetrock but must be the correct output.

Stage 1 - Planning and Design: An accurate heat loss and design must be done to compute the amount of film to use. The R-value of floor coverings must be considered in the design.

Stage 2 - Ground Work: Run any underground electric wiring and required service connections.

Stage 3 - Prior to Electric Radiant Installation: Make sure joist bays are clean and clear of sharp objects that could damage film.

Stage 4 - During Electric Radiant Installation: Make certain correct amount of film is installed with no folds and proper clearances.

Stage 5 - Prior to Cover: Inspect the integrity of the film and that all connections are intact. Perform continuity checks on all circuits.

Stage 6 - Trim Out: Install thermostats and other controls. Protect film from abuse that could damage it.

Stage 7 - Start-up: Test all sequences of operation.

Stage 8 - Floor Covering: Since the output of systems installed below the subfloor is limited, use as conductive a floor coverings as possible. When installed in ceilings, check with manufacturer when using thick sheetrock or other coverings.

Figure 9-21 Thin Electric with Plastic Film

9

9•4 Summary

Choosing the correct hydronic heating appliance requires matching the operating needs of the system with the operating ranges and efficiencies of the choices. Most radiant heating systems are installed with boilers since they can provide a wide range of temperatures. They are very durable and, when mated with an indirect water tank (a water tank with a heat exchanger), they can provide large amounts of domestic hot water from the same appliance. Often the boiler can run other hydronic options such as hot water baseboard, or be converted to forced air by a coil in an air handler. In choosing the best appliance, the hydronic designer frequently is looking at a whole heating and domestic water system, not just radiant. Snow melting and pool heating are sometimes added as well.

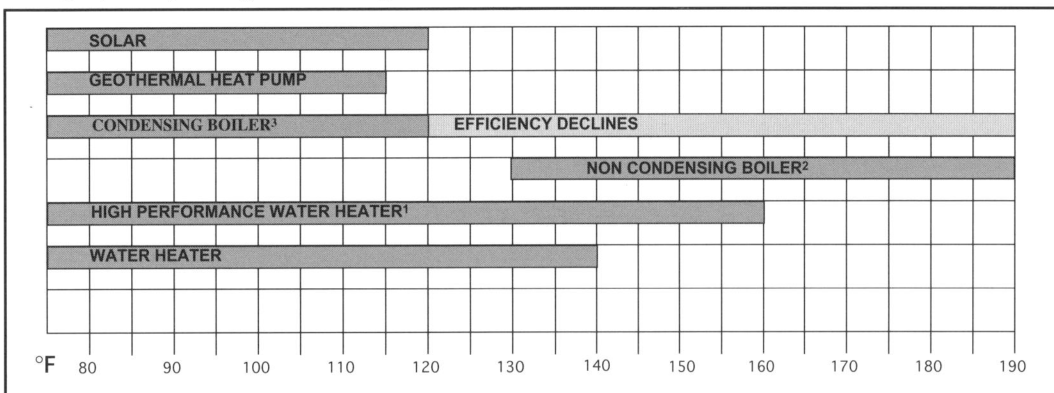

Typical Temperature Operating Range of Hydronic Heat Sources

Notes:

1) Most water heaters have 140 °F as maximum set point, some allow for higher settings up to 160 °F or 180°F.

2) Most non-condensing boilers need to operate in the 150 °F to 190 °F range to prevent condensation in the flue or boiler sections. Mixing strategies are employed when using lower temperature radiant systems.

3) Condensing boilers typically extract and extra 9% of efficiency in the boiler by condensing water vapor

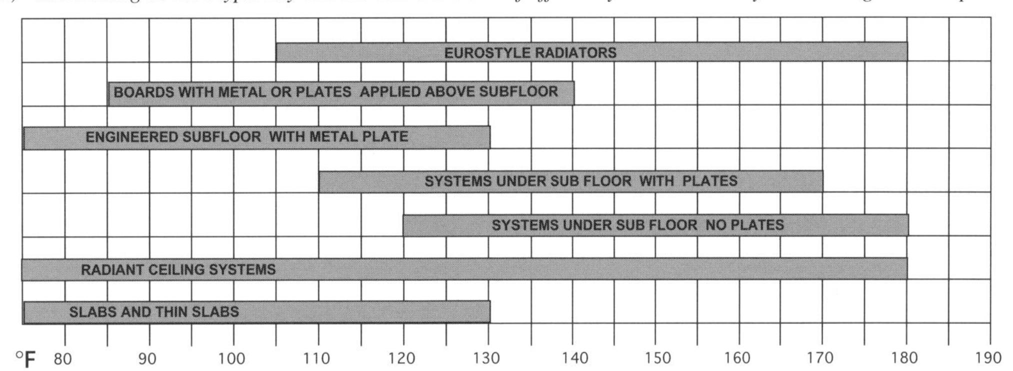

Typical Supply Water Temperature Range For Different Hydronic Radiant Heating Systems

Notes:

1) Electric cable systems often operate in the same temperature range where applicable.

2) These are typical operating ranges for these systems and are intended for conceptual overview only.

3) Hydronic ceiling systems with tubing typically operate in the lower range of temperature shown.

Figure 9-22 Temperature Ranges for Heat Sources and Systems

index